TEST YOUR HIGHER CHEMISTRY

(Revised Higher)

by

J.J. MacDonald

ISBN 0 7169 3192 3

ROBERT GIBSON · Publisher
17 Fitzroy Place, Glasgow, G3 7SF.

INTRODUCTION

The questions in this book cover the revised *Higher Chemistry* syllabus for the *Scottish Certificate of Education*. About a third of the questions cover work needed to gain a C in the examination. The rest of the questions, marked by a margin line, cover work for a higher award: A or B. Some questions test knowledge and understanding. Others test problem solving.

Answers to all questions are given at the end of the book. Additional information and explanations are also given where these may be helpful.

The Data Book referred to in some questions is issued by the *Scottish Examination Board* and may be purchased from *Robert Gibson and Sons, Glasgow.*

The problems and solutions make considerable use of international symbols and abbreviations. These are given on page 3 for reference.

CONTENTS

Questions and answers by J. J. MacDonald, B.Sc., A.R.C.S.T., M.Sc..

SYMBOLS

A	ampere
A_r	relative atomic mass
(aq)	dissolved in water
atm	atmosphere (101 325 Pa)
bar	standard pressure (100 000 Pa)
c	concentration
[A]	concentration of A
c	specific heat capacity
C	coulomb
°C	degree Celsius
cm	centimetre
ΔH	enthalpy change
ΔT	temperature change
d	day
dm^3	cubic decimetre
e	charge of electron[1]
e or e^-	electron
E_A	activation energy
E_i	ionisation energy
E_k	kinetic energy
E_p	potential energy
$E^{\ominus}$	standard electrode potential
F	Faraday constant
g	gram
(g)	gaseous
I	electric current
J	joule
K	kelvin
kg	kilogram
kJ	kilojoule
L	litre
L	Avogadro constant[3]
m	metre
M	molar = mol L^{-1}
m	mass
min	minute
mL	millilitre
mm	millimetre
M_m	molar mass[4]
mol	mole
M_r	relative molecular mass
n	amount
n	number of electrons
n	neutron
N_A	Avogadro number[5]
nm	nanometre
p	proton
Q	electric charge
Q_m	molar charge[4]
r	rate of reaction
s	second
(s)	solid
t	tonne (1000 kg)
t	time
$t_{1/2}$	half-life period
T	temperature
u	atomic mass unit[6]
V	volt
V	volume
V_m	molar volume[4]
z	charge number of ion

1. e = 1.602×10^{-19} C per electron or per proton.
2. F = 96 500 C mol^{-1} (of any singly charged species).
3. L = 6.02×10^{23} (particles) mol^{-1} (of substance).
4. In these terms, 'molar' means 'per mole'.
5. N_A = 6.02×10^{23}, a number only. (L = N_A (particles) mol^{-1} (of substance).)
6. $^1/_{12}$th of the mass of a C-12 atom.

ABBREVIATIONS

a.m.	atomic mass
a.m.u.	atomic mass unit
A.N.	atomic number
b.p.	boiling point
e.m.f.	electromotive force
f.m.	formula mass
m.m.	molecular mass
M.N.	mass number
m.p.	melting point
r.a.m.	relative atomic mass
r.f.m.	relative formula mass
r.m.m.	relative molecular mass
r.t.p.	room temperature and pressure
r.t.s.p.	reference temperature and standard pressure (25 °C and 1 bar)

UNIT 1

CONTROLLING REACTION RATES

1. Hydrogen peroxide decomposes according to the equation:

$$H_2O_2(aq) \rightarrow H_2O(l) + O_2(g)$$

The volume of oxygen evolved from 200 mL of solution (to which a catalyst had just been added) was measured against time. The results are shown in the table.

Time /s	Volume of O_2 /mL
0	0
20	30
40	42
60	51
80	54
100	55

(a) Draw a graph of volume of O_2 against time. (2)
(b) Estimate the time taken for complete decomposition. (1)
(c) Calculate the average rate of reaction between
(i) 20 and 30 s; (2)
(ii) 40 and 50 s. (2)
(d) Explain the difference between the rates in (c)(i) and (c)(ii). (1)

2. Make a labelled drawing of the apparatus that could be used for the rates experiment of Question 1. (2)

3. In a rates experiment, the concentration of iodine was measured as the reaction progressed. Data are shown opposite.

Time /min	Concentration of iodine, $[I_2]$ /mol L^{-1}
0	2.0×10^{-5}
4	1.3×10^{-5}
9	7.0×10^{-6}
16	1.1×10^{-6}
18	2.5×10^{-6}
23	5.0×10^{-7}
29	Zero

(a) (i) Is iodine a reactant or a product in this reaction? (1)
(ii) Explain your answer to (i). (1)
(b) The quantity *rate* is expressed in units that depend on the experiment. In this experiment, concentration is measured against time. State the rate unit to be used here. (1)
(c) Draw a graph of concentration of iodine against time. (2)
(d) Calculate the average rate of reaction between 0 and 5 min. (2)
(e) A mistake has been made in measuring a concentration. Use your graph to identify the mistake and say what the concentration should have been. (1)

4. (a) What is a variable? (1)
(b) Name four variables that control reaction rates. (2)

5. Name the type of substance that increases the speed of a reaction and is left over after the reaction. (1)

6. A quantity of hydrochloric acid and excess marble chips (calcium carbonate) were placed on a balance linked to a computer. The acid was added to the marble. The decrease in mass of the reaction mixture with time was displayed as a graph, which is shown below.

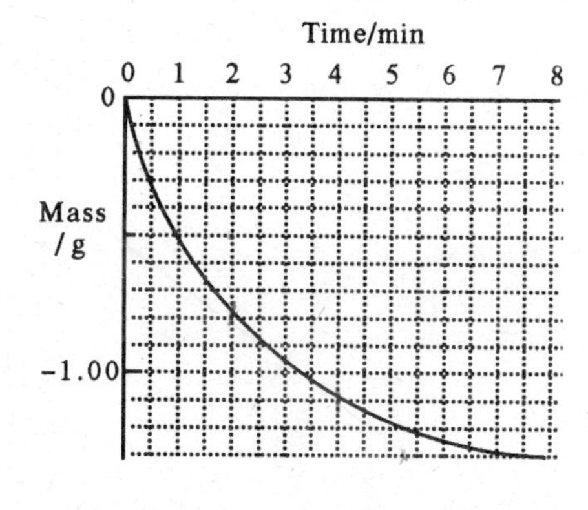

(a) Identify
    (i) the independent variable, (1)
    (ii) the dependent variable. (1)
(b) Explain these terms. (1)
(c) Calculate the rate of reaction at 3 min. (2)
(d) Write a balanced chemical equation for the reaction. (1)
(e) Calculate the mass of chalk that reacted with the acid. (2)

7. Identical pieces of magnesium were dropped into equal volumes of hydrochloric acid of different concentrations (c). The time (t) for complete reaction was noted in each case.

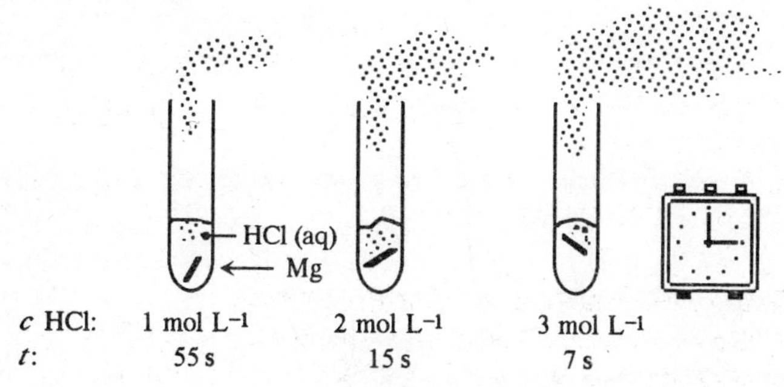

(a) What is the relation between rate of reaction (r) and time (t)? (1)
(b) Calculate the average rate of reaction for each concentration of acid. (3)
(c) Draw (i) a t/c curve and (ii) a r/c curve for this reaction. (4)
(d) Use the t/c curve to predict the time for reaction with 1.5 mol L^{-1} HCl. (1)

8. Name (i) a very slow reaction, (ii) a very fast reaction. (1)

9. The rates of three reactions were plotted against the concentrations of reactants A (first reaction), B (second reaction) and C (third reaction).

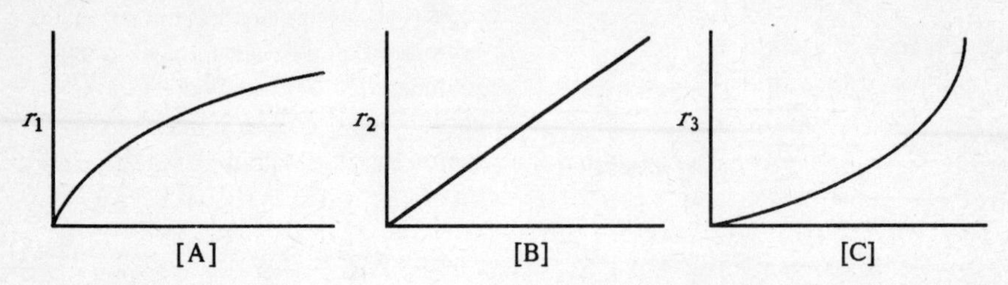

(a) What in general is the relation between rate of reaction and concentration of reactant? (1)

(b) In which reaction is rate directly proportional to concentration of reactant? (1)

10. Sodium thiosulphate in acid solution decomposes according to the equation:

$$Na_2S_2O_3(aq) + 2HCl(aq) \rightarrow 2NaCl(aq) + SO_2(g) + S(s)$$

Three experiments were carried out using quantities as shown in the table. These quantities were added to the same volume of excess hydrochloric acid (180 mL).

Expt.	0.1 mol L^{-1} $Na_2S_2O_3$ /mL	Water /mL
1	20	0
2	15	5
3	10	10

↑ Light transmitted
A B
3
2
1
0 10 20 30 40
t/s

The reaction mixture was transparent at first and then became cloudy (due to the formation of colloidal sulphur).

The progress of each reaction was followed by computer using a detector for transmitted light. Three curves were obtained.

(a) In which experiment is the initial reaction fastest? (1)

(b) Explain why you have chosen your answer to (a). (1)

(c) Which experiment takes the longest time? (1)

(d) Explain why the answers to (a) and (c) are not contradictory. (1)

Line AB cuts the curves at times when the mass of sulphur in suspension is the same in each experiment.

(e) Complete the following table with respect to line AB.

Expt.	Concentration of $Na_2S_2O_3$ /mol L^{-1}	Duration of reaction /s	Average initial rate /s^{-1}
1	0.01	t_1	r_1
2	c_2	5	r_2
3	c_3	t_3	0.083

(6)

(f) Draw a graph of average initial rate against concentration. (2)

11. (a) Which variable was under test in the experiments of Question 10? (1)
(b) (i) Name another variable that was actively kept constant. (1)
(ii) Explain how it was kept constant. (1)

12.

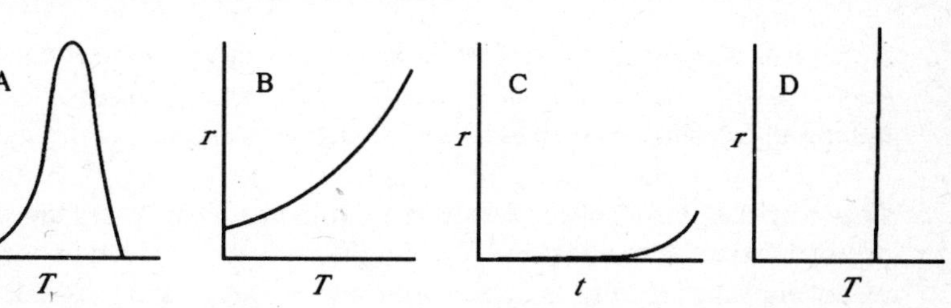

A, B and D are r/T curves; C is a r/t curve. Identify the curve that represents

(a) the explosive detonation of a mixture of methane and air; (1)
(b) the action of an enzyme; (1)
(c) the reaction between sodium thiosulphate and hydrochloric acid; (1)
(d) a reaction with an incubation period. (1)

13. The reaction between hydrogen and iodine to give hydrogen iodide is reversible: the reaction can occur both forwards and backwards.

$$H_2(g) + I_2(g) \rightleftharpoons 2HI(g)$$

The theory of reaction requires that molecules must collide in order to react. Assume that the only particles in the reaction mixture are H_2, I_2 and HI molecules.

(a) How many kinds of collision (H_2 with H_2, H_2 with I_2, and so on) occur in the reaction mixture? (1)
(b) (i) Of these collisions, which may produce H_2 and I_2 molecules? (1)
(ii) Draw a diagram to show how bond rearrangement might occur in (i). (1)
(c) In addition to colliding, what two other conditions must be satisfied if two molecules are to react? (2)

14. Ammonium nitrite is unstable and decomposes according to the equation

$$NH_4NO_2 \rightarrow 2H_2O(l) + N_2(g)$$

The volume of nitrogen evolved from each of four solutions of ammonium nitrite was measured over the same period of time. The nitrogen in each experiment was collected at the same room temperature.

Experiment	Solution Volume /mL	Solution Concentration /mol L^{-1}	Temperature /°C
1	20	0.5	50
2	20	0.25	50
3	40	0.5	50
4	20	0.5	60

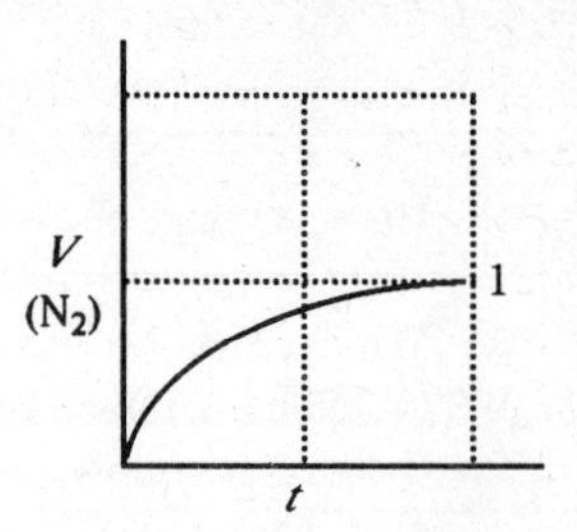

A graph of volume of N_2 against time for experiment 1 is shown above. Copy graph 1 (with the reference lines). Draw curves for experiments 2, 3 and 4 on this copy, labelling the curves 2, 3 and 4 respectively. (3)

15. Cast-iron contains about 20% carbon by volume (about 4% by weight), most of which exists as uncombined graphite flakes.

Cast-iron cannon-balls recovered from sunken wrecks and dried out have been known to burst into flames. Explain. (2)

16.

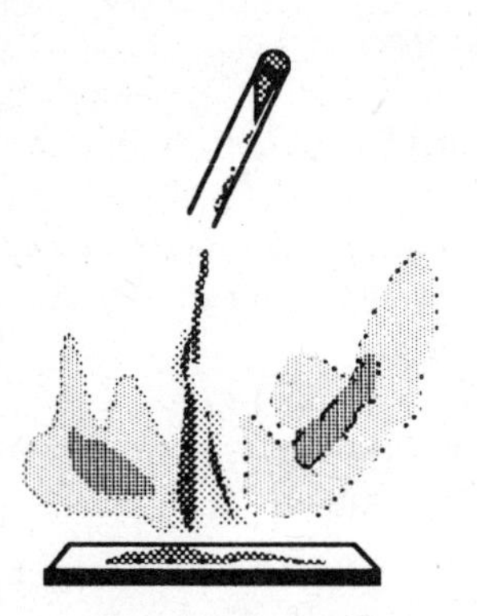

Pyrophoric powders are powders that catch fire on exposure to air. Pyrophoric lead is made by roasting lead tartrate out of contact with air. When the lead cools and is tipped out of the tube it immediately bursts into flame.

(a) What is the physical property of pyrophoric lead that causes it to behave like this? (1)

(b) Why does this physical property produce this result? (2)

17. White phosphorus dissolves in carbon disulphide, a liquid that evaporates quickly (a volatile liquid). When the solution is poured over cotton wool and left for a few minutes, the wool suddenly bursts into flame. Offer a reason for this and write an equation for the reaction. (3)

18. The graph shown opposite was obtained by plotting the loss in mass for a reaction between 5 g marble chips and excess hydrochloric acid.
    Copy the graph and, on the same graph, draw the curve that would be obtained when 2.5 g of powdered marble are added to the same volume of hydrochloric acid. (1)

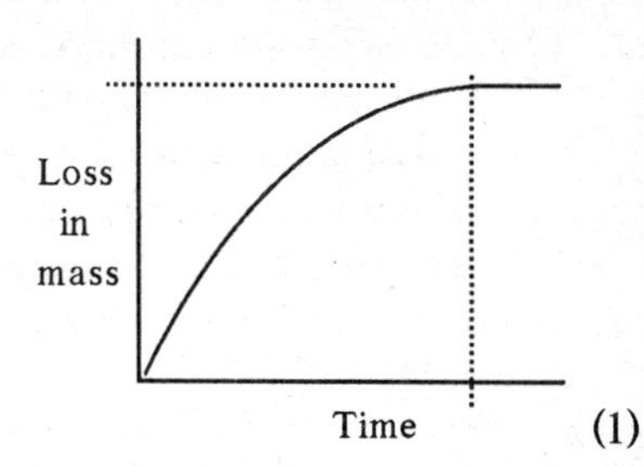

19. (a) Calculate the surface area of a cube of 1 cm edge (a cm cube). (1)
    (b) (i) Calculate the volume of a cube of 1 nm (10^{-9} m) edge (a nm cube). (1)
    (ii) Calculate the number of nm cubes in one cm cube. (1)
    (c) (i) Calculate the surface area of a nm cube. (1)
    (ii) Calculate the total surface area of the number of nm cubes in (b)(ii). (1)
    (d) Explain why a powder is more reactive than a lump. (1)

20. In 'clock reactions' the end of the reaction is signalled by the sudden appearance of a colour or a precipitate. In one clock reaction, the appearance of a blue colour was timed (using the same quantities) at different temperatures. Draw a t/T graph and use it to find the temperature rise (ΔT) that doubles the rate of reaction. (3)

T/°C	t/s
20	117
30	59
40	30
50	17
60	8

21. Hydrogen peroxide solution decomposes slowly over long periods according to the equation: $H_2O_2(aq) \rightarrow H_2O(l) + \frac{1}{2}O_2(g)$; $\Delta H = -98$ kJ mol^{-1}.
    The activation energy for the decomposition is $E_A = 40$ kJ mol^{-1}.
    Catalysts accelerate the decomposition. Some catalysts are shown in the table together with their activation energies.

Catalyst	E_A/kJ mol^{-1}
Dust	36
Manganese dioxide, MnO_2	29
Iron(II) ion, Fe^{2+}	26
Platinum, Pt	25
Catalase (an enzyme)	11

(a) Is the decomposition of H_2O_2 endothermic or exothermic? (1)
(b) Why is the decomposition slow in the absence of a catalyst? (1)
(c) Which is the most effective catalyst? (1)
(d) Draw a spike graph of E_A/catalyst. (2)
(e) Draw an enthalpy diagram for the uncatalysed reaction and the platinum-catalysed reaction. (The diagram need not be to scale.) (3)

(f) The Bombardier Beetle uses the explosive decomposition of hydrogen peroxide when mixed with catalase to fire a high-velocity corrosive spray at a potential predator. The beetle allows five seconds between each shot. Give two reasons for this. (2)

22. The graph shows the distribution of speeds among hydrogen (H_2) molecules at a temperature T_1.

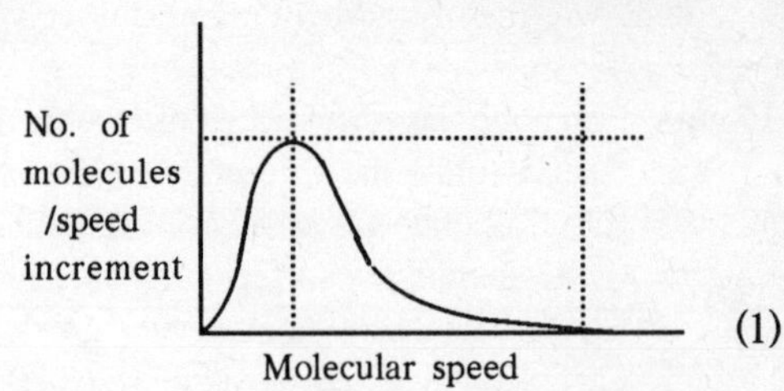

(a) What does the area under the curve represent? (1)

Copy the curve and the reference lines. On this copy, draw the distribution for the same number of

(b) H_2 molecules at a higher temperature T_2. (1)
(c) N_2 molecules at T_1. (1)
(d) Explain your answers to (b) and (c). (2)

23. The graph shows the distribution of kinetic energies among reactant molecules at T_1. (Note that this curve has a different shape from that in question 22.) Copy the curve and the reference lines.

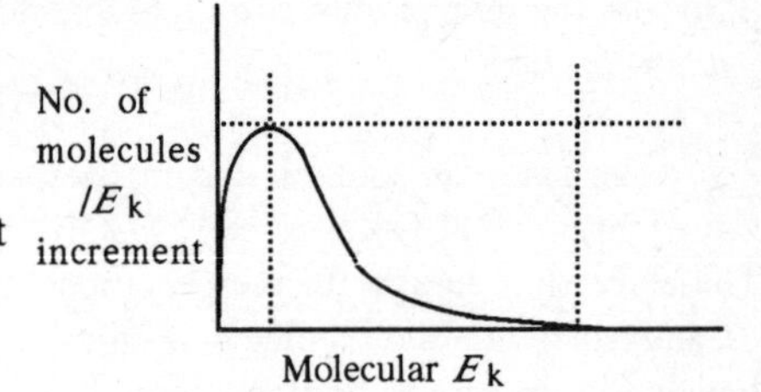

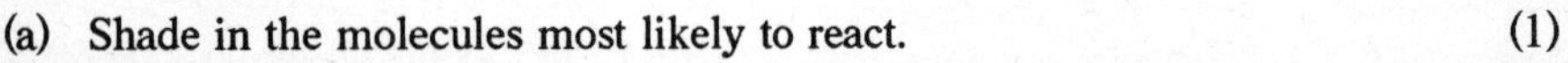

(a) Shade in the molecules most likely to react. (1)
(b) (i) Considering your answer to (a), mark E_A for this reaction. (1)
(ii) Define E_A in this context. (1)
(c) (i) Draw a distribution curve for a higher temperature T_2. (1)
(ii) A small temperature rise has a marked effect on reaction rate. Explain. (1)
(d) How does a catalyst affect
(i) the shape of the curve at T_1; (1)
(ii) the value of E_A; (1)
(iii) the shaded area at T_1? (1)

24. In the manufacture of nitric acid, platinum, in the form of a gauze, is used as the catalyst. After a time the gauze develops thick patches and thin patches.

(a) Why is the catalyst in the form of a gauze? (1)
(b) Why does the gauze develop thick and thin patches? (1)
(c) Sketch a possible shape taken by the activated complex when A_2 and B_2 molecules react to give AB molecules. (2)

25. The graph shows the curve obtained by plotting the volume of hydrogen liberated when 1 mole of magnesium powder reacts with a large excess of 0.1 mol L^{-1} hydrochloric acid at r.t.s.p.. From the grid select the curves that are correct for each of the rate experiments (a) to (f).

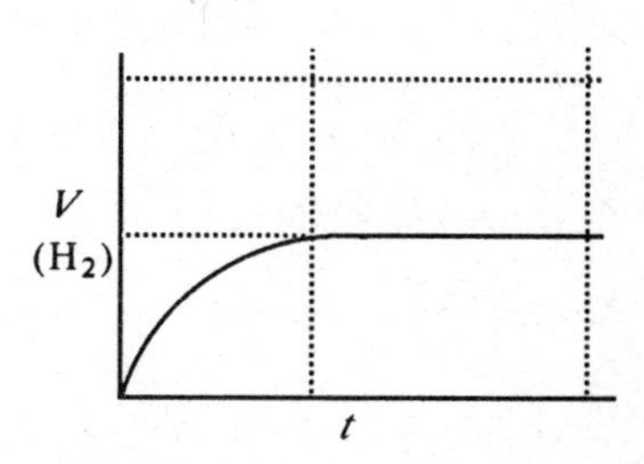

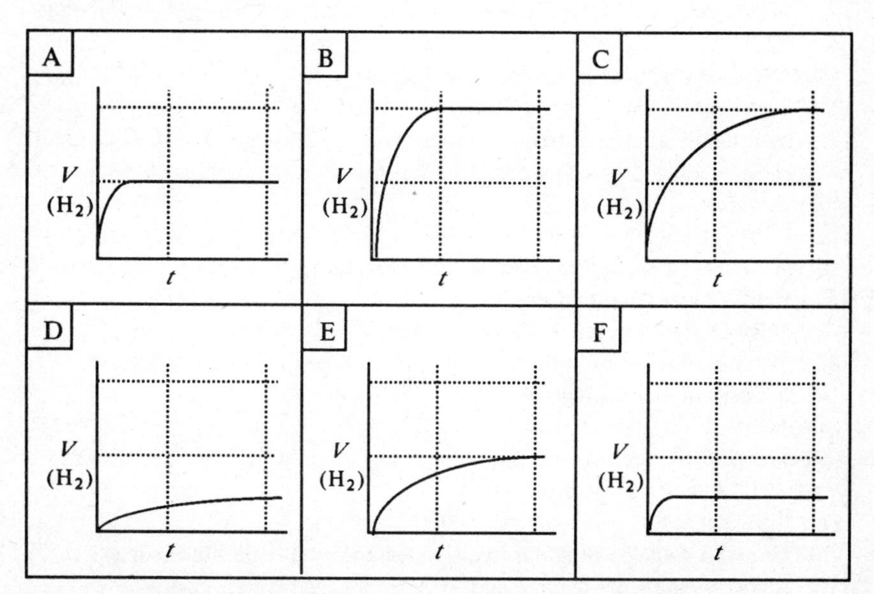

(a) 1 mol of zinc powder instead of 1 mol of magnesium powder. (1)
(b) 2 mol of magnesium powder instead of 1 mol of magnesium powder (but acid still in excess). (1)
(c) 1 mol of sodium instead of 1 mol of magnesium. (1)
(d) 1 mol of magnesium turnings instead of 1 mol magnesium powder. (1)
(e) 1 mol of magnesium powder but acid at 50 °C instead of the reference temperature of 25 °C (but gas still collected at r.t.s.p.). (1)
(f) 1 mol of magnesium powder with 0.05 mol L^{-1} hydrochloric acid instead of 0.1 mol L^{-1} acid (but acid still in excess). (1)

26. Enthalpy diagrams for two reactions, A → B and P → Q, are shown below.

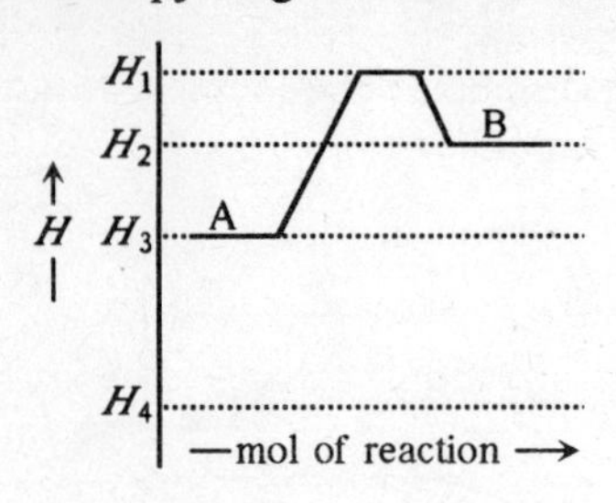

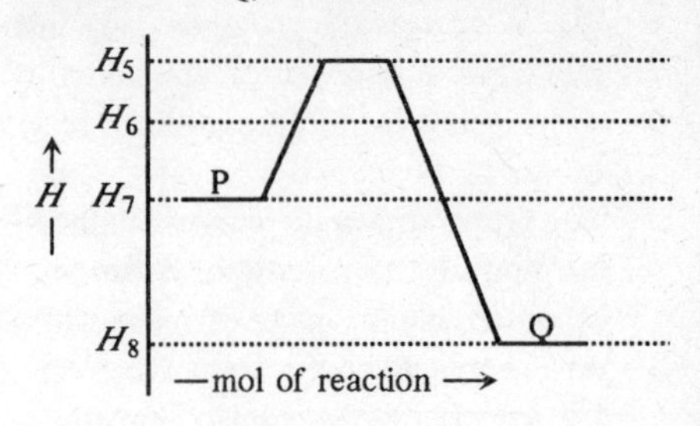

(a) Write expressions for
(i) ΔH (A → B); (ii) ΔH (P → Q); (iii) E_A (A → B); (iv) E_A (P → Q). (4)

(b) Is ΔH (A → B) (i) positive or negative; (ii) endothermic or exothermic? (1)

(c) Write expressions for (i) E_A (B → A); (ii) E_A (Q → P). (2)

(d) Define E_A as used in an enthalpy diagram. (1)

27. Constant temperature is a basic assumption of enthalpy diagrams for chemical reactions: products lose or gain heat so that reactants and products have the same temperature.

(a) (i) Is ΔH a change in potential or kinetic energy? (1)
(ii) Explain your answer. (1)

(b) What is the usual reference temperature? (1)

(c) When a reaction mixture cools during a reaction, is the reaction exothermic or endothermic? (1)

28. E_A and ΔH for the reaction $HgO(s) \rightarrow Hg(l) + \frac{1}{2}O_2(g)$ are 99 kJ mol^{-1} and +91 kJ mol^{-1} respectively.

(a) Draw an enthalpy diagram (which need not be to scale) for the reaction. (2)

(b) State the value of (i) ΔH and (ii) E_A for the reverse reaction. (2)

29. From the information in question 28, state (i) ΔH and (ii) E_A for the reaction

$$2HgO(s) \rightarrow 2Hg(l) + O_2(g)$$ (2)

30. E_A and ΔH for the reaction $N_2(g) + 3H_2(g) \rightarrow 2NH_3(g)$ are 236 kJ mol^{-1} and −94 kJ mol^{-1} respectively.

(a) What is E_A for the reverse reaction. (1)

(b) What effect does a catalyst have on (i) E_A; (ii) ΔH? (2)

31. Define the term 'activated complex'? (1)

32. Say whether the following statements about a catalyst are true or false. It

(a) must be a solid. (1)
(b) is unchanged at the end of the reaction. (1)
(c) must possess a large surface area. (1)
(d) provides an alternative route of lower activation energy. (1)
(e) supplies energy to the reactants. (1)
(f) is not involved in the reaction. (1)
(g) always speeds up a reaction. (1)
(h) can be radiation in the form of light. (1)

33. A potential energy diagram for the decomposition of $H_2O_2(aq)$ is shown below.

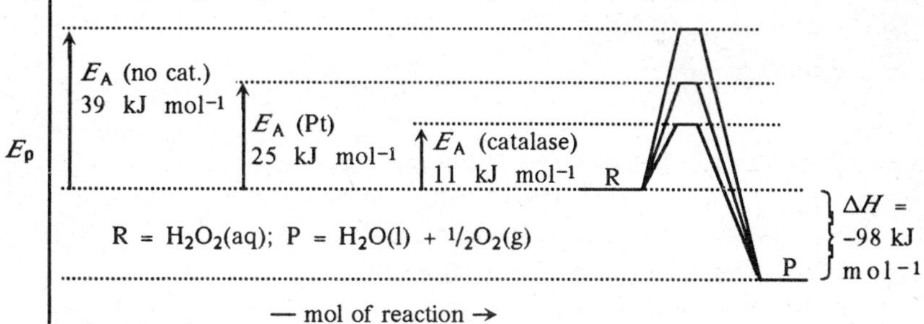

(a) Which reaction has the lowest activation energy? (1)
(b) Which reaction would be fastest? (1)
(c) What effect does the introduction of a catalyst have on
    (i) the rate of the reaction; (1)
    (ii) the enthalpy change? (1)
(d) State the value of E_A for the reverse Pt catalysed reaction. (1)
(e) Is the catalysis with Pt homogeneous or heterogeneous? (1)

34. State whether the catalyses shown in the grid are homogenous, colloidal or heterogeneous.

A $N_2O(g) \rightarrow N_2(g) + {}^1/_2O_2(g)$; catalyst = $Cl_2(g)$.	B $N_2(g) + 3H_2(g) \rightarrow 2NH_3(g)$; catalyst = Fe(s).
C $C_6H_{12}O_6(aq) \rightarrow 2C_2H_5OH(aq) + 2CO_2(g)$; catalyst = zymase.	D $H_2O_2(aq) \rightarrow H_2O(l) + {}^1/_2O_2(g)$; catalyst = $MnO_2(s)$.

(4)

35. Mineral (inorganic) catalysts and organic enzymes are sensitive substances. Very small amounts of arsenic, mercury and lead can destroy the catalysts; carbon monoxide and cyanide ion are lethal to enzymes.

(a) (i) State the name given to substances like arsenic. (1)
(ii) Explain their action on catalysts. (1)
(b) (i) State the name given to substances like cyanide ion. (1)
(ii) Explain their action on enzymes. (1)

36. Aluminium powder is stable and does not react with oxygen; the rate at which aluminium oxide forms is virtually zero.
A mixture of aluminium powder and iodine placed on a gauze smoulders and then bursts into flame: aluminium oxide forms rapidly, liberating clouds of purple iodine vapour. (The reaction can be initiated by a drop of water.)

(a) What is the function of iodine in this reaction? (1)
(b) (i) Given that aluminium first combines with iodine to form aluminium iodide and that the iodide then combines with oxygen to form the oxide, write equations for the two reactions involved. (2)
(ii) Use these equations to support your answer in (a). (2)
(iii) Write a net equation for the reaction. (1)
(c) A drop of water may be used to initiate the reaction. Name three other ways by which a reaction may be initiated. (3)

37. When barium chloride solution is added to dilute sulphuric acid a white precipitate forms immediately.

(a) Why is this reaction so fast? (1)
(b) Write an equation to support your answer in (a). (1)

38. Derive an expression that links forward activation energy, $E_{A,f}$, enthalpy change, ΔH, and reverse activation energy, $E_{A,r}$. (1)

39. What effect does the addition of an inhibitor to an enzyme-catalysed reaction have on

(a) the activation energy of the reaction; (1)
(b) the enthalpy change for the reaction? (1)

UNIT 2
FEEDSTOCKS AND FUELS

The chart on pages 32 and 33 may help you to answer some questions in this Unit.

1. Explain the terms (a) raw material, (b) finite resource, (c) feedstock, (d) consumer material, (e) fuel. (5)

2. The structural formulae of some organic products are shown below.

$C_2H_5OC_2H_5$
Ether, an anaesthetic.

$[-CH_2-C(CH_3)=CH-CH_2-]_n$
Rubber: a polymer

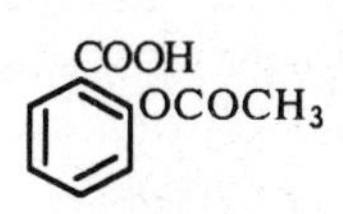

Aspirin: an analgesic

COOH
Benzoic acid; a food preservative.

$C_6H_5CH_2CONHCH-C-S-C(CH_3)_2$, $CO-N-CH-COOH$
Penicillin: an antibiotic.

$CH_3-C(CH_3)=CH-CH_2-CH_2-C(CH_3)=CH-CH_2OH$
Geraniol: a component of rose oil perfume.

$H_2N(CH_2)_nNH_2$
For putrescine, n = 4, for cadaverine, n = 5; two of the foulest smelling substances known.

$NH_2CH(CH_2COOH)-CONH-CH(CH_2C_6H_5)-COOCH_3$
Aspartame; an artificial sweetener.

Benzopyrene; a chemical that causes cancer; found in tobacco smoke

(The hexagon symbol stands for six carbon atoms and a variable number of hydrogen atoms.)

(a) Name two elements that are common to all these compounds. (1)
(b) Three other elements occur to a lesser extent. Name these in decreasing frequency of occurrence. (1)
(c) Which of these compounds has (i) the smallest, (ii) the largest mass per molecule? (1)
(d) Calculate the molecular mass of putrescine. (1)

3. "Raw materials are complex materials that are broken down into feedstocks. Feedstocks are composed of small molecules. These are built up into consumer materials composed of large molecules." Quote one sequence in support of this statement. (2)

4. Select compounds whose structural formulae are shown in question 2 to answer this question.

   (a) Name two hydrocarbons. (1)
   (b) (i) Name two unsaturated compounds. (1)
   (ii) Explain why they are unsaturated. (1)
   (c) Write full structural formulae for the hydroxyl, carboxyl and amino groups. In each case name a compound that contains the group. (3)
   (d) What are groups like this called? (1)
   (e) Name a compound that has two of the same groups per molecule. (1)

5. Write full structural formulae for
   (a) 2-methylpropene. (1)
   (b) 2-methylbut-1-ene. (1)
   (c) 2-methylbut-2-ene. (1)
   (d) 1-chloro-2-methylpropane. (1)
   (e) (i) cis-1,2-dichloroethene; (ii) trans-1,2-dichloroethene. (2)
   (f) Pent-2-ene. (1)

6. Write full structural formulae for the six isomers of C_4H_8. (6)

7. Name the substances whose structural formulae are shown below.

```
(a)      CH3        CH3            (b)          CH3 CH3
         |          |                           |   |
     CH3-CH-CH2-CH-CH3                CH3-CH2-CH-CH-CH3

(c)             CH3                (d)    CH3-C-C2H5
                |                             ||
     CH3-CH2-C=CH2                           CH2
```

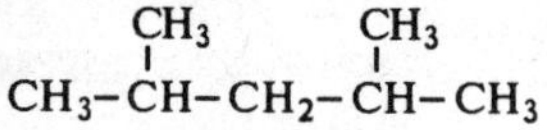

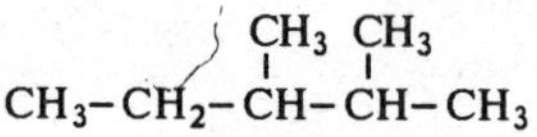

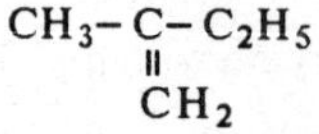

(e) $CH_3C(CH_3)_2CH_2CH(CH_3)CH_3$ (f) $CH_2{=}C(CH_3)CH_2CH_3$ (6)

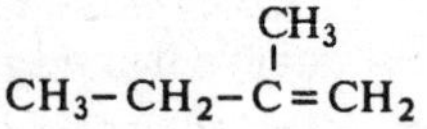

8. Write skeletal structural formulae for the nine isomers of heptane, C_7H_{16}. (Skeletal formulae show only the carbon backbone of the molecule.) (4)

9. Write skeletal structural formulae for the 18 isomers of octane, C_8H_{18}. (Number correct 18-17: superhuman; 16-14: excellent; 13-12: very good; 11-9: good; 8-6: fair; <6: poor.) (-)

10. Name the substances whose structural formulae are shown below.

```
(a)  H  H  H        (b)  H  H  H        (c)  H    O        (d)  H    H
     |  |  |             |  |  |             |   //             |   /
   H-C--C--C-OH        H-C--C--C-H         H-C-C              H-C-N
     |  |  |             |  |  |             |   \              |   \
     H  H  H             H  OH H             H    OH            H    H
```

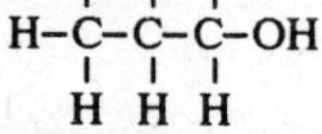

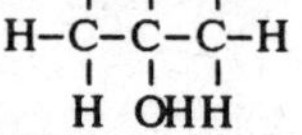

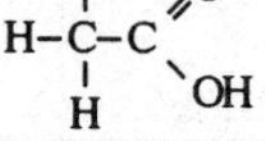

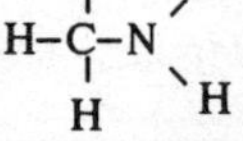

(4)

11. Write solutions for (a) to (j).

	Pure ethanol	Pure ethanoic acid	
Short structural formula	CH_3CH_2OH	(a)	(1)
Picture of the molecule	(b)		(1)
Molar mass/g mol^{-1}	46	(c)	(1)
Flammability at r.t.	(d)	Burns if sustained	(1)
pH of 1 mol L^{-1} solution	(e)	Approx. 3	(1)
Reaction with potassium	H_2 evolved	(f)	(1)
Reaction with magnesium	(g)	H_2 evolved	(1)
Raw material used to make it	(h) (renewable)	Crude oil (finite)	(1)
How it is made	Fermentation	(i)	(1)
Consumer product	Beers, wines, spirits	(j)	(1)

12. Natural polymerisation links monomers –head–tail–head–tail–head–tail–. Synthetic polymerisation often links monomers –head–head–tail–tail–head–head–.
   (a) (i) Which way will the monomer $NH_2CHXCOOH$ polymerise? (1)
   (ii) Explain your answer to (i). (1)
   (iii) Name a natural class of polymer formed from a monomer of this kind. (1)
   (iv) Draw a short structural formula of a portion of the polymer molecule showing three monomer residues. (1)
   (b) (i) Explain why $NH_2(CH_2)_6NH_2$ and $HOOC(CH_2)_4COOH$ polymerise –head–head–tail–tail–. (1)
   (ii) Suggest a reason why industry prefers this method of polymerisation. (1)
   (iii) Name a synthetic class of polymer made from monomers of this kind. (1)
   (iv) Draw a short structural formula of a portion of the polymer molecule. (1)

13. **Petroleum Gases**

Oil from the Brent system contains lots of gas. This is piped ashore. When cooled under pressure, ethane, propane and butane condense leaving methane as a gas.

Methane (NG)

Ethane
Propane
Butane
(NGLs)

   (a) State the principal use of methane. (1)
   (b) Methane is transported as a liquid, LNG.
   (i) Refer to a data book and state the temperature in °C to which methane must be cooled to liquefy it. (1)
   (ii) What does 'LNG' stand for? (1)
   (c) (i) What does 'NGL' stand for? (1)
   (ii) How are the NGLs separated? (1)
   (d) State the principal use of ethane. (1)
   (e) Propane and butane are bottled as LPG.
   (i) What does 'LPG' stand for? (1)
   (ii) Give a use for propane and for butane. (2)

14. ***Cracking ethane.***
Ethane from natural gas and natural gas liquids is cracked to give ethene. The yield is high, about 80%. Side reactions (polymerisations) give 20% heavier fractions. These consist of about 5% propene, 10% fuel gases and 5% gasoline.
Cracking propane.
When propane (same source) is cracked, the yield of propene is less than 80%.

(a) What is the meaning of 'cracking'? (1)
(b) Construct a pie chart for the products of cracking of ethane. (2)
(c) (i) Write a balanced equation for the polymerisation of ethene to octene. (1)
(ii) Name the reaction that occurs when octene changes into octane? (1)
(iii) Write a chemical equation for the reaction in (ii). (1)
(d) Why is the yield of propene from propane less? (1)
(e) Write an equation (short structural formulae) for the cracking of propane. (1)

15. ***Separating natural gas liquids.***
The graph shows how the boiling points of ethane, propane and butane vary with pressure.

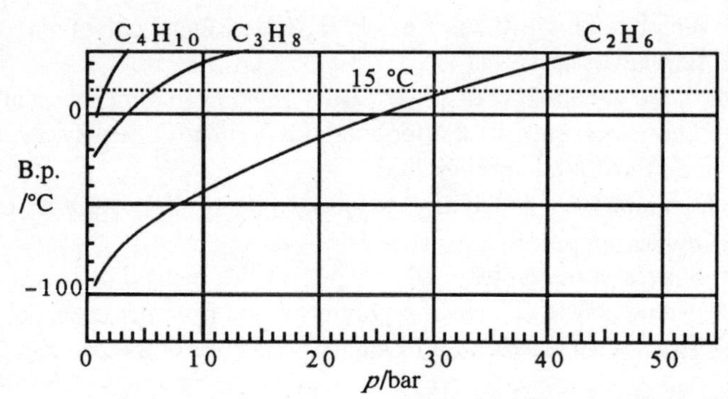

Liquids may be fractionally distilled because they boil at different temperatures at the same pressure. They may also be fractionally distilled because they boil at different pressures at the same temperature. NGLs (liquefied ethane, propane and butane) are separated in this way.

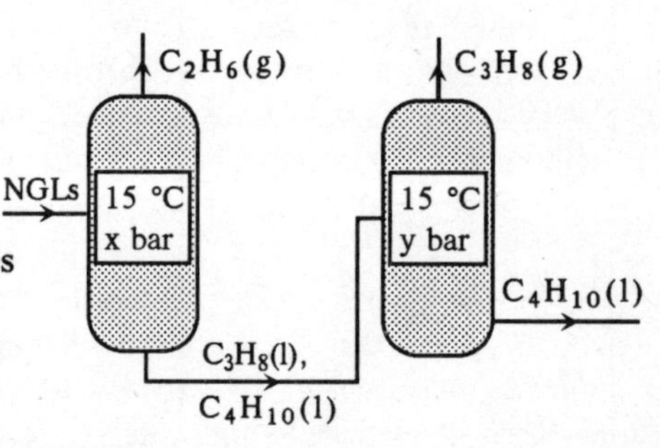

(a) Use the graph to estimate the boiling point of ethane at 1 bar. (1)

(Contd.)

(b) Is ethane a liquid or a gas at –50 °C and 20 bar? (1)

(c) Why is the b.p. of propane less than that of butane at the same pressure. (2)

(d) Estimate values for x and y that would allow separation of ethane, propane and butane as shown in the diagram. (2)

16. Construct a summary flow diagram for the materials and reactions described in Questions 13, 14 and 15. (3)

17. Steam cracking, hydrocracking and cat-cracking are variants of thermal cracking. Explain why (i) steam, (ii) hydrogen and (iii) catalysts are used. (4)

18. Naphtha is a mixture of hydrocarbons from C_6-C_{10}. It contains straight alkanes, branched alkanes, cycloalkanes (naphthenes) and aromatics.

Naphtha is further separated into light naphtha, C_6-C_8, average b.p. about 100 °C, and heavy naphtha, C_8-C_{10}, average b.p. about 150 °C, and processed as shown below.

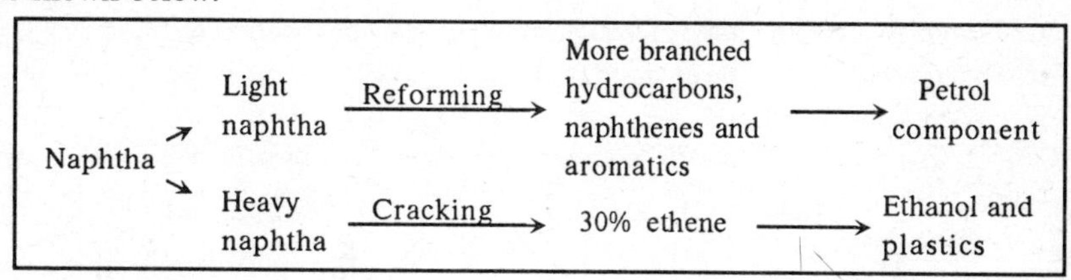

For light naphtha

(a) Draw and name a full structural formula for

(i) the 'heaviest' straight alkane; (1)

(ii) the lightest branched alkane; (1)

(iii) the lightest cycloalkane; (1)

(iv) the simplest aromatic. (1)

(b) Which of the four kinds of hydrocarbon is most abundant? (1)

(c) (i) What is meant by 'reforming'? (1)

(ii) Use full structural formulae to illustrate your answer to (i) by reforming a molecule of octane. (1)

(iii) Why is reforming used to make petrol components? (1)

For heavy naphtha

(d) Why does heavy naphtha have a higher average b.p. than light naphtha? (1)

(e) Why is the yield of ethene much less than that from ethane (80%)? (1)

(f) Ethene is reacted with steam to give ethanol.

(i) Name the type of reaction. (1)

(ii) Write an equation for the reaction. (1)

19. Define the term 'functional group'. (1)

20. Two ways of writing the structural formula of benzene are shown below.

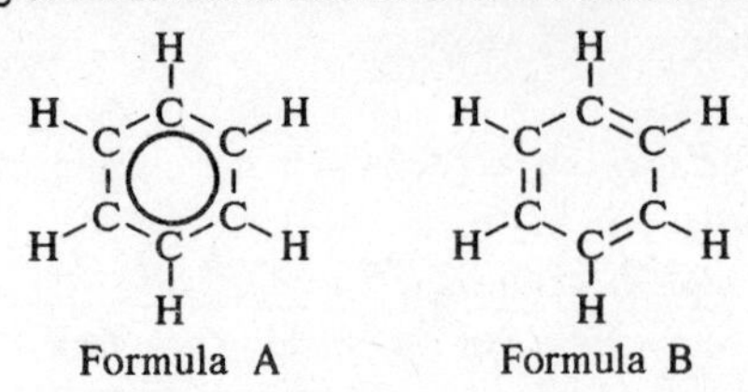

Formula A   Formula B

(a) (i) Explain why formula A is generally preferred to formula B. (1)
    (ii) Quote one piece of experimental evidence in support of this view. (1)
(b) Write ring symbols for each of these formulae. (1)

21. Destructive distillation of coal gives four main products, one of which is coal tar.

(a) Name two terms either of which may replace 'destructive distillation'. (1)
(b) What are the other three main products? (1)

Coal tar is rich in aromatic feedstocks. Formulae for some of these are shown below.

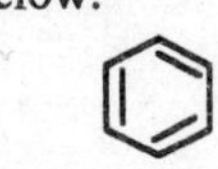

Benzene(l), C_6H_6

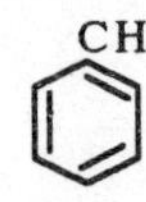

Toluene(l)

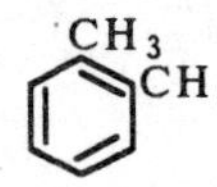

Xylene(l)

Pyridine(l)

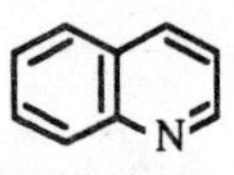

Quinoline(l),

OH

Phenol(s)

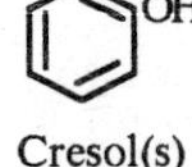

Cresol(s)

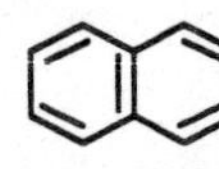

Naphthalene(s)

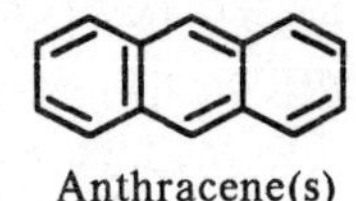

Anthracene(s)

(c) Write molecular formulae for each of these except benzene. (4)
(d) Coal tar used to be the major source of aromatics but has been replaced in recent years by an alternative source. Name this source. (1)

22. An incomplete reaction is shown below.

$$+ \; Br_2 \longrightarrow \; ?$$

(a) Name the organic reactant. (1)
(b) (i) Write the ring formula of the product; (1)
    (ii) Name the product; (1)
    (iii) Write the molecular formula of the product. (1)

23. An incomplete reaction is shown below.

$$+ \; 2Br_2 \longrightarrow \; ?$$

Cyclohexa-1,3-diene

(a) Write the ring formula of the product. (1)
(b) Name the product. (1)
(c) Write the molecular formula of the product. (1)

24. An incomplete reaction is shown below.

$$+ \; Br_2 \longrightarrow \; ?$$

(a) (i) Name the organic reactant. (1)
  (ii) Write the molecular formula of the reactant. (1)
(b) (i) Write the ring formula of the organic product. (1)
  (ii) Name the products. (1)
  (iii) Write the molecular formula of the organic product. (1)

25. Identify the types of reaction occurring in Questions 22, 23 and 24. (3)

26.

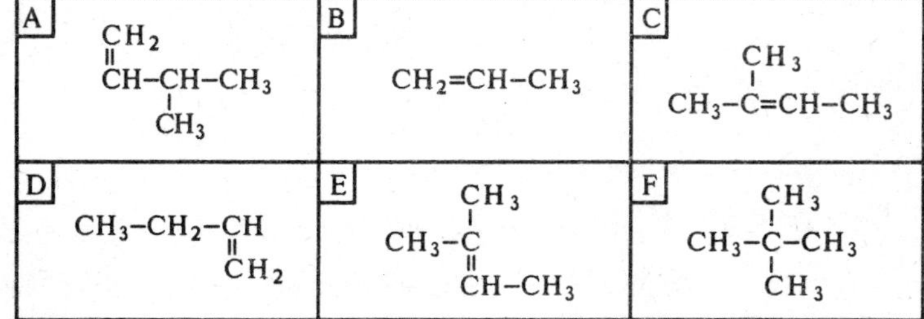

In the above grid, identify
(a) the alkane. (1)
(b) an isomer of but-2-ene. (1)
(c) two hydrocarbons that are the same. (1)
(d) 3-methylbut-1-ene. (1)
(e) the monomer used to make polypropene. (1)

27. Write systematic and trivial names for

(a) $CH_2{=}CHCl$. (2)
(b) $C_6H_5CH_3$. (2)
(c) C_6H_5COOH. (2)

28. Name the compound formed when bromine reacts with

(a) but-1-ene. (1)
(b) but-2-ene. (1)
(c) 2-methylpropene. (1)

29.

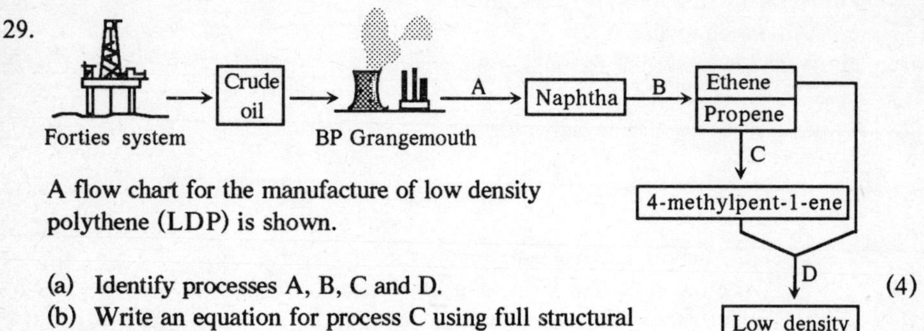

A flow chart for the manufacture of low density polythene (LDP) is shown.

(a) Identify processes A, B, C and D. (4)
(b) Write an equation for process C using full structural formulae. (2)
(c) Name the two monomers used to make LDP. (2)

This polythene has a low density because side chains prevent the polymer molecules coming close together. The side chains are due to the inclusion of 4-methylpent-1-ene.

(d) (i) Using short structural formulae, show a portion of the polymer chain derived from the monomer sequence:
+ propene + 4-methylpent-1-ene + propene +. (2)
(ii) Draw a circle round the side-chain. (1)
(iii) Name the side chain. (1)

30. Coal varies a lot in composition. The older it is, the more carbon it contains and the higher is its calorific value, C.V., (energy per gram).
Lignite is a young coal. It contains 60% carbon and has a C.V. of 20 kJ g^{-1}. (The difference is volatile matter, minerals and moisture.) Bituminous coal is middle-aged. It contains 80% carbon and has a C.V. of 32 kJ g^{-1}. Anthracite is old. It has 90% carbon. Its C.V. is 34 kJ g^{-1}.

(a) Display the above information in a table. (3)
(b) Which mineral in coal is responsible for acid rain. (1)
(c) Name one substance found in the 'volatile matter'. (1)

31. The traditional way of making *ethanol* uses yeast to ferment *sugars*. An *enzyme* in yeast converts the sugars into ethanol. Explain the words in italic. (3)

32. Name and write formulae for the compounds formed when but-1-ene is hydrated. (2)

33. Organic compounds are classified as shown below.

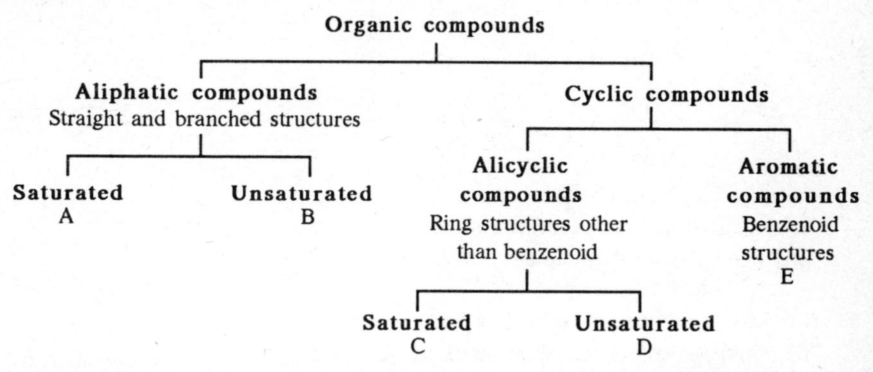

Place the following compounds in their correct category (A, B, C, D or E):

(a) Cyclohexane.
(b) $CH_2{=}CHCH_3$.
(c) Toluene.
(d) $CH_3CH_2CH(CH_3)CH_2CH_3$.
(e) Propanoic acid.
(f) C_2H_5OH.
(g) Cyclopentene.
(h) C_6H_5Br (phenyl bromide). (8)

34. When ethanol is heated in the apparatus shown below, a reaction occurs and a gas collects in the receiver.

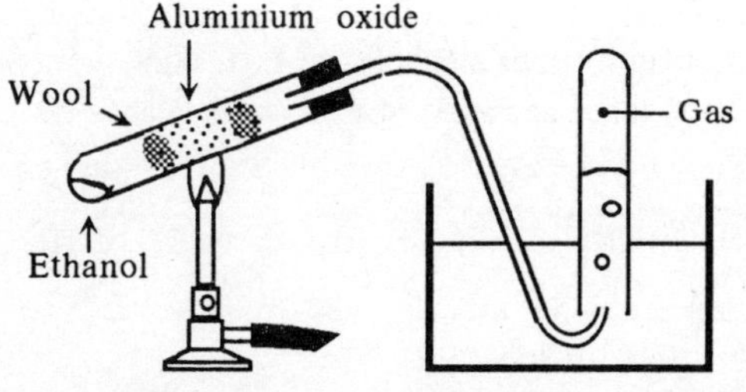

(a) (i) State the type of reaction. (1)
(ii) Write an equation for the reaction. (1)
(b) What is the purpose of the aluminium oxide? (1)
(c) Name the gas. (1)
(d) Name a test for the gas. (1)
(e) (i) When ethanol is replaced with propan-1-ol, a different gas is obtained. Name the gas. (1)
(ii) Write an equation with full structural formulae for the reaction. (2)

35.

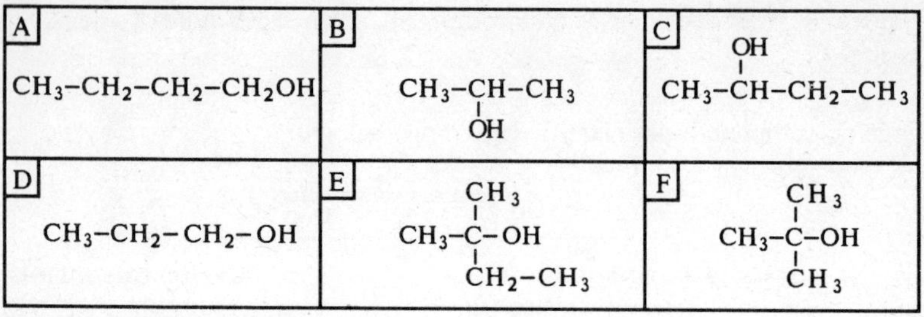

A	B	C
$CH_3-CH_2-CH_2-CH_2OH$	$CH_3-CH-CH_3$ $\vert$ OH	OH $\vert$ $CH_3-CH-CH_2-CH_3$
D	**E**	**F**
$CH_3-CH_2-CH_2-OH$	CH_3 $\vert$ CH_3-C-OH $\vert$ CH_2-CH_3	CH_3 $\vert$ CH_3-C-OH $\vert$ CH_3

Closed questions

(a) Which alcohol is an isomer of D? (1)

(b) Identify 2-methylbutan-2-ol. (1)

(c) Which alcohol can be dehydrated to

(i) but-1-ene only; (1)

(ii) a mixture of but-1-ene and but-2-ene? (1)

(d) Which alcohol can be oxidised to

(i) propanal; (1)

(ii) propanoic acid; (1)

(iii) propanone? (1)

Open questions

(a) State which alcohols are (i) primary, (ii) secondary.

(b) Select the isomer(s) of A.

(c) Which alcohol(s) can be dehydrated to methylpropene?

36. Give systematic names for the alcohols whose short formulae are shown below.

(a) 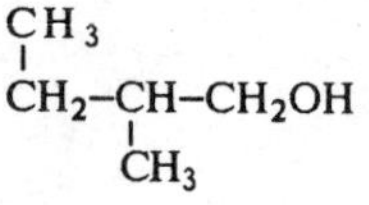
CH_3
$\vert$
$CH_2-CH-CH_2OH$
$\vert$
CH_3

(b) 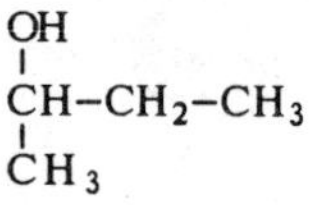
OH
$\vert$
$CH-CH_2-CH_3$
$\vert$
CH_3

(c) 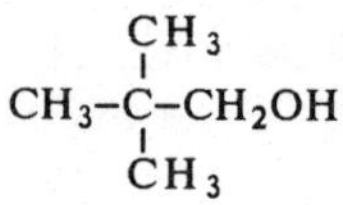
CH_3
$\vert$
CH_3-C-CH_2OH
$\vert$
CH_3

(3)

37. In an investigation a student had to identify four alcohols, A, B, C and D, whose formulae were known to be those shown below.

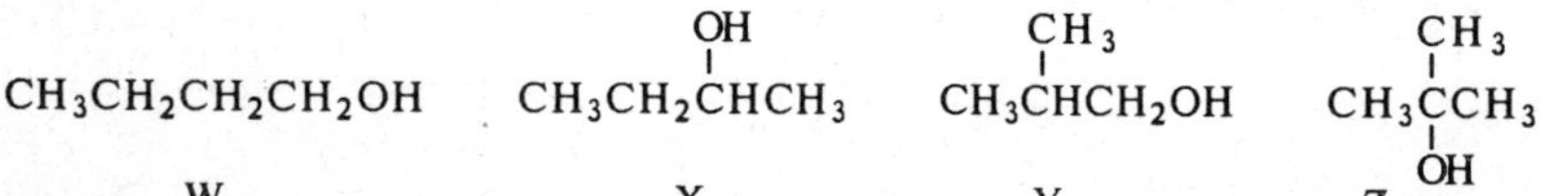

W	X	Y	Z
$CH_3CH_2CH_2CH_2OH$	OH $\vert$ $CH_3CH_2CHCH_3$	CH_3 $\vert$ CH_3CHCH_2OH	CH_3 $\vert$ CH_3CCH_3 $\vert$ OH

First: 5 mL of each alcohol were placed in separate tubes in a warm bath. Five drops of acidified potassium dichromate solution were added to each. A, C and D each gave a positive colour change. There was no colour change with B. This identified B.

(Contd.)

Second: 5 mL of A, C and D were placed in each of three test tubes in the water bath. Five drops of acidified potassium dichromate solution were added to each. A and D developed similar ethanal-like smells. C developed a smell like that of acetone (propanone). This identified C.

Third: The boiling point of A was found to be 118 °C. D boiled at 108 °C. Reference to a data book identified A. The remaining alcohol, therefore, was D.

(a) (i) Identify B. (1)
(ii) What is meant by "positive colour change"? (1)
(iii) Explain your choice for B. (2)

(b) (i) Identify C. (1)
(ii) Explain your choice for C. (2)

(c) (i) Identify A and D. (1)
(ii) Explain your choice for A. (1)
(iii) Draw an apparatus used to find the b.p. of the alcohols. (2)
(iv) Explain why D has a lower b.p. than A. (1)

(d) Give systematic names for W, X, Y and Z. (4)

38. Explain why glucose, $C_5H_{11}O_5CHO$, and sucrose, $C_6H_{11}O_5{-}O{-}C_6H_{11}O_5$, may be distinguished by testing with Fehling's (Benedict's) Reagent. (2)

39. Why does methanoic acid, HCOOH, give a positive result with Fehling's Reagent? (2)

40.

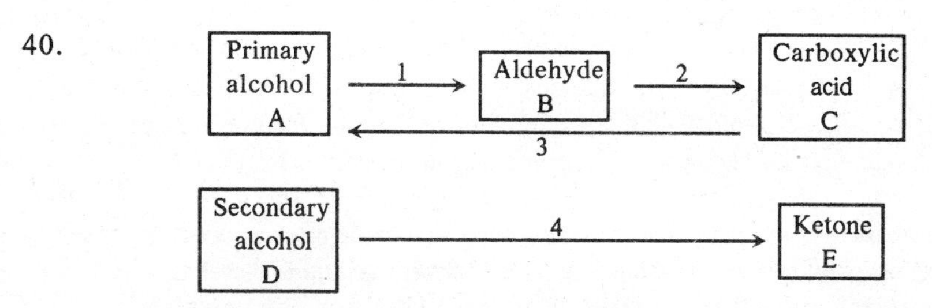

The following questions refer to the above chart.

(a) Link the letters A to E with the formulae:
(i) CH_3COOH;
(ii) $CH_3CHOHCH_3$;
(iii) CH_3CH_2OH;
(iv) CH_3COCH_3;
(v) CH_3CHO. (3)

(Contd.)

(b) (i) State which of the oxidations, 1, 2 or 4, occurs most easily. (1)
(ii) Explain your choice. (2)
(c) Write an equation for reaction 4. (1)

41. Name the ketones (i) $CH_3CH_2COCH_3$, (ii) $CH_3CH(CH_3)COCH_3$. (2)

42. Write the molecular formula for 3-methylcyclohexanone, (ring with =O and CH_3).

Questions 43 and 44 refer to the following chart/table.

Raw Materials		Feedstocks		Consumer Products
A →	Coke —1→			
B →	Naphtha —2→	**Synthesis gas**	→ Methanol ···→	D
C →	Methane —3→			

43. Synthesis gas (syngas) is made by steam reforming feedstocks such as coke, naphtha and methane (reactions 1, 2 and 3). The feedstocks are derived from raw materials.

(a) Name the raw materials A, B and C. (1)
(b) What is 'steam reforming'? (1)
(c) Name the two gases in synthesis gas. (1)

44. Naphtha is a complex mixture of alkanes, cycloalkanes and aromatics in the range C_6 to C_{10}. If we use C_8H_{16} as an average formula for the fraction, the chemical equation for steam reforming of naphtha (reaction 2) is then:

$$C_8H_{16}(g) + 8H_2O(g) \rightarrow \underline{8}CO(g) + \underline{16}H_2(g)$$

This equation tells us that 8 volumes of carbon monoxide are produced for every 16 volumes of hydrogen. Synthesis gas made from naphtha is therefore a mixture by volume of 8 parts carbon monoxide to 16 parts hydrogen; the volume ratio of $CO:H_2 = 8:16 = 1:2$.

(a) Write balanced equations for steam reforming reactions 1 and 3. (2)
(b) From these equations, state the volume ratio of $CO:H_2$ for synthesis gas made from (i) coke, (ii) methane. (1)
(c) Which synthesis gas has a composition most suitable for making methanol:

$$CO(g) + 2H_2(g) \rightarrow CH_3OH(l)$$ (1)

(d) Suggest a name for D. (1)

45. ***Gasification of coal.*** Coal is gasified (steam reformed) by heating with steam and oxygen. When the amounts of steam and oxygen are equal (i.e., when one molecule of H_2O is consumed for one molecule of O_2), simultaneous chemical equations for the process are:

1. $C(s) + H_2O(g) \rightarrow CO(g) + H_2(g)$
2. $2C(s) + O_2(g) \rightarrow 2CO(g)$

Reaction 1 is endothermic; reaction 2 is exothermic.

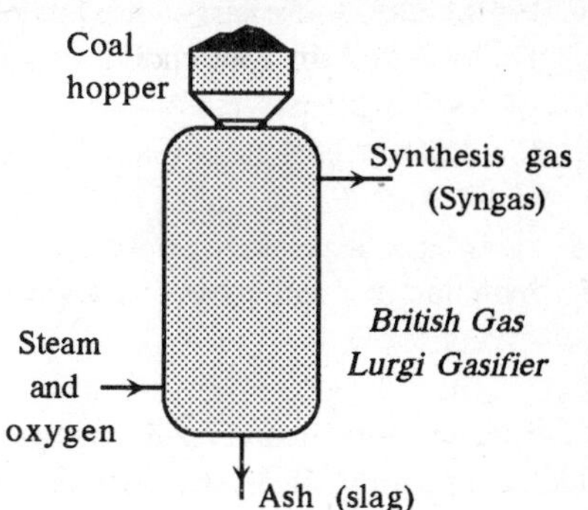

(a) Define the terms endothermic and exothermic. (1)

(b) State an advantage of a process in which endothermic and exothermic reactions are run at the same time. (1)

(c) What effect does the use of oxygen in addition to steam have on the composition of the synthesis gas? (1)

(d) State the volume ratio of $CO{:}H_2$ of the syngas from the Lurgi process. (1)

(e) Construct a bar graph that shows the composition of syngas from coke, naphtha, methane (Question 44, page 26) and coal gasification. (3)

46. The main products from the cracking of 1 tonne each of naphtha and ethane are shown in the bar graph.

Ethene sells at £320, propene at £250, fuel gases at £105 and gasoline at £100, all per tonne.

Naphtha and ethane feedstocks cost £65 and £40 per tonne, respectively.

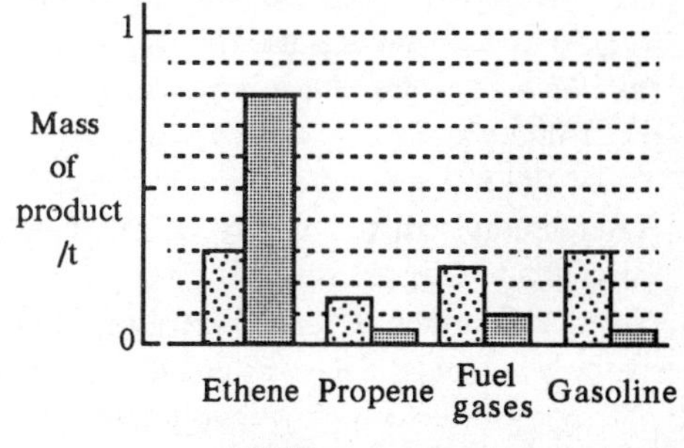

(a) Calculate the income per tonne of
(i) cracked naphtha; (2)
(ii) cracked ethane. (2)

(b) Calculate the profit for (a)(i) and (ii). (2)

(c) Calculate the profit per tonne of ethene made from
(i) naphtha; (1)
(ii) ethane. (1)

(d) Which is the more profitable way of making ethene? (1)

(e) The cost of feedstocks, however, is not the only cost. List four other costs. (2)

47. As well as ethene (and other products), the cracking of naphtha gives benzene. Benzene and ethene are then used to make an addition polymer (plastic).
    (a) Name the polymer. (1)
    (b) Draw a portion of the polymer molecule showing three repeating units. (1)

48. In order to survive, a manufacturer must make a profit. However, a responsible manufacturer must also consider factors not directly concerned with profit. Name two of these. (1)

49. Bacteria may contaminate wine and oxidise the alcohol to acid. This sours the wine and changes it into vinegar.
    The deliberate souring of wine is the traditional way of making vinegar. A large wooden cask, pierced with holes, contains beech shavings inoculated with special bacteria. Wine trickles over the shavings at about 35 °C and is oxidised to vinegar.

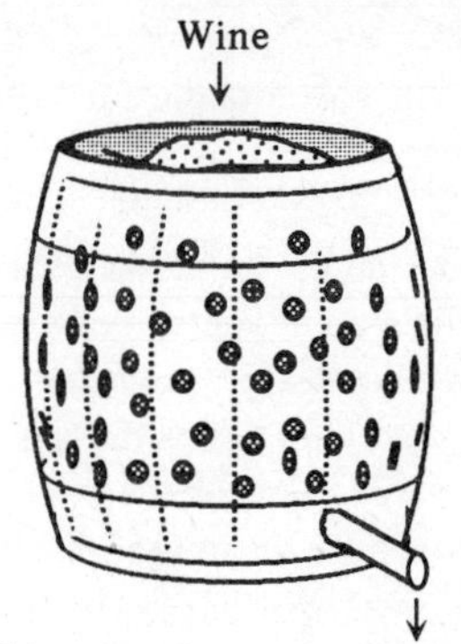

    (a) Name (i) the alcohol in wine and (ii) the acid in vinegar. (1)
    (b) What purpose do the holes in the cask serve? (1)
    (c) Why are shavings used? (1)
    (d) Using full formulae, write an equation for the reaction in the cask. (2)

50. Draw the full structural formula of a carboxylic acid that is an isomer of the ester methyl ethanoate, CH_3COOCH_3. Name the acid. (2)

51. Draw full structural formulae for the simplest monoprotic and diprotic carboxylic acids. Give systematic and trivial (traditional) names for each. (6)

52. Name the acids that have the formulae
    (i) $CH_3CH_2CH(CH_3)COOH$; (1)
    (ii) $CH_3CH(CH_3)CH_2CH_2COOH$. (1)

53. (a) Use the chart on pages 32-33 to complete the table shown below.

Processes used to make petrol and diesel

Petrol		Diesel	
Physical	Chemical	Physical	Chemical

(3)

    (b) Why is petrol a more expensive fuel than diesel? (1)

54. ***Cracking of wax distillate***

Wax distillate is a product of the vacuum distillation of the residue from the primary distillation of crude oil. It is a mixture of saturated hydrocarbons averaging about $C_{30}H_{62}$.

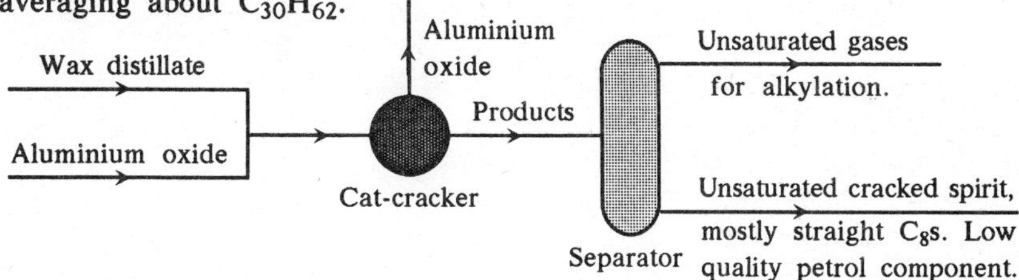

(a) What is the function of aluminium oxide? (1)

(b) In the interests of economy, what is done with the aluminium oxide leaving the cat-cracker? (1)

(c) A $C_{30}H_{62}$ molecule is cracked into two octene molecules, a heptane molecule, a butene molecule and one other molecule.
(i) Name the other molecule. (1)
(ii) Draw its full structural formula. (1)
(iii) Name a consumer material made from molecules of (i). (1)

Alkylation is a reaction in which small molecules react to give branched petrol molecules. A typical alkylation is shown below.

$$CH_3CH(CH_3)CH_3 + CH_2{=}CHCH_2CH_3 \longrightarrow CH_3-\underset{\displaystyle CH_3}{\overset{\displaystyle CH_3}{CH}}\!\!-CH_2-\underset{\displaystyle CH_3}{\overset{\displaystyle CH_3}{C}}-CH_3$$

Methylpropane from primary distillation	But-1-ene from cat-cracking of wax distillate	A typical alkylate: high quality petrol component

(d) Give the systematic name of the alkylate shown above. (1)

(e) Write a short formula for an alkylate that is even more highly branched. (1)

(f) Explain the difference in quality of cracked spirit and alkylate for use as petrol components. (2)

55. Petrol and diesel fuels are complex mixtures of straight, branched, cyclic and aromatic hydrocarbons and other substances.

(a) Use the chart on pages 32-33 and the information in Question 54 above to make a list of the components of each fuel. (4)

(b) In winter, as much as 14% butane may be added to petrol to make it more volatile. Explain. (2)

(c) (i) Which material is added to diesel oil in winter? (1)
(ii) Why is it added? (1)

56. Internal combustion engines depend on the ignition of fuel and air in a cyclinder (usually four per engine). Petrol and diesel engines use different methods of ignition and different kinds of fuel.

    (a) Contrast the methods of ignition in each kind of engine. (3)
    (b) State the average number of carbon atoms per molecule of (i) petrol and (ii) diesel. (1)
    (c) Which molecular structure of a fuel is best for
    (i) a petrol engine; (1)
    (ii) a diesel engine? (1)

57. **Sewage gas**

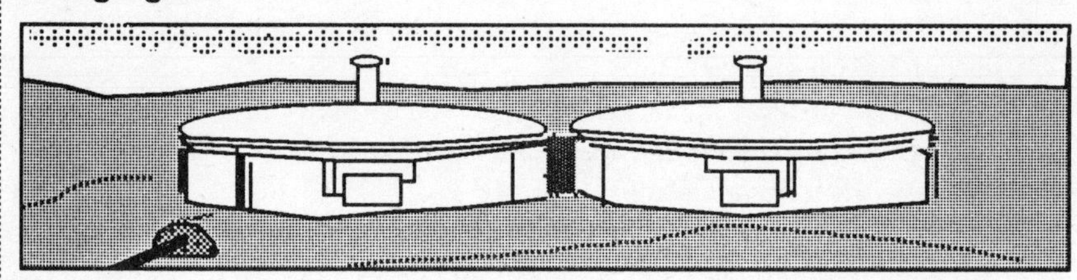

Sewage digesters

"The biogas scheme at the Finham Sewage Treatment Plant came on line in July 1992. The scheme, operated by Biogas Generation Ltd., is contracted to supply 975 kW of electrical power. The electricity is generated using three spark ignition engines fuelled by biogas produced from the digestion of sewage sludge." *Review, Quarterly Magazine of Renewable Energy, Issue 20.*

(a) What is the principal gas in biogas? (1)
(b) How is digestion achieved? (1)
(c) State two reasons why air must not be allowed to enter the digesters. (2)
(d) Suggest an economic use for the waste heat from the spark ignition engines. (1)
(e) Sketch and label a small laboratory-scale digester. (3)

58. Amines are nitrogen-containing compounds produced by rotting biomass. They react (and smell) like ammonia.

    (a) Describe a test (other than smell) for an amine. (2)
    (b) Write the name and full structural formula for the simplest amine. (1)
    (c) Ammonia reacts with water according to the equation:

$$NH_3 + H_2O \rightarrow NH_4^+ + OH^-$$

Write the analogous equation for the reaction of $C_2H_5NH_2$ with water and name the positive ion produced. (2)

59. Condensation of urea, $(NH_2)_2CO$, with methanal (formaldehyde), HCHO, gives a thermosetting plastic.

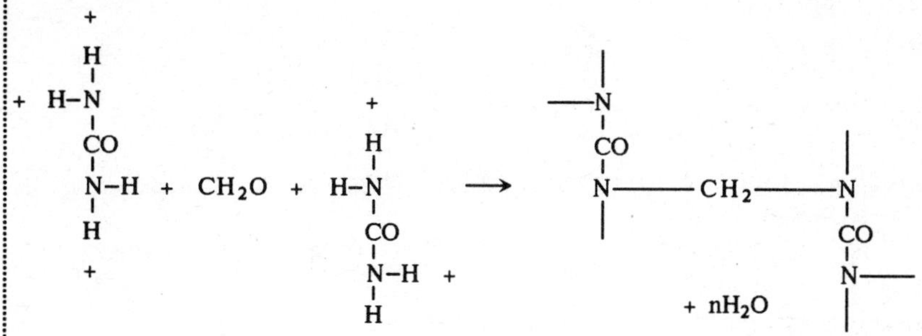

```
          +
          H
          |
     + H–N                   +               —N
          |                                    |
          CO                 H                 CO                |
          |                  |                 |                 |
          N–H  +  CH2O  +  H–N       ⟶        N———CH2———N
          |                  |                 |                 |
          H                  CO                                  CO
                             |                                   |
          +                  N–H  +                              N—
                             |
                             H                       + nH2O      |
                             +                  Urea-methanal plastic
```

(a) What does 'thermosetting' mean? (1)

(b) Why is urea-methanal a thermosetting plastic? (1)

(c) In view of your answer to (b),

  (i) what is wrong with the way the reaction is drawn above; (1)

  (ii) how might the drawing be improved? (2)

(d) State one physical property of this plastic and a corresponding use. (1)

The first synthetic thermosetting plastic was made by Baekeland in 1907 by condensing phenol with methanal. It is called bakelite.

```
       +
       H
       |                       OH
 + H–[C6H3]–H  +  CH2O  +  H–[C6H3]–H  +   ⟶   Bakelite  +  nH2O
       OH                      |
     Phenol                    H
                               +
```

(e) Draw the structural formula of bakelite using the equation for urea-methanal as a guide. (2)

60. Classify each of the organic compounds whose formulae are shown below.

(i) CH_3CHO; (ii) $CH_3CH_2CH_3$; (iii) $C_2H_5COCH_3$; (iv) C_3H_7COOH; (v) CH_2OHCH_3; (vi) $CH_2{=}CHCH_3$; (vii) C_6H_{12} (satd.); (viii) HCOOH. (4)

The author is indebted to the Understanding British Industry Project sponsored by the CBI Education Foundation, which provided the information for some questions in this Unit.

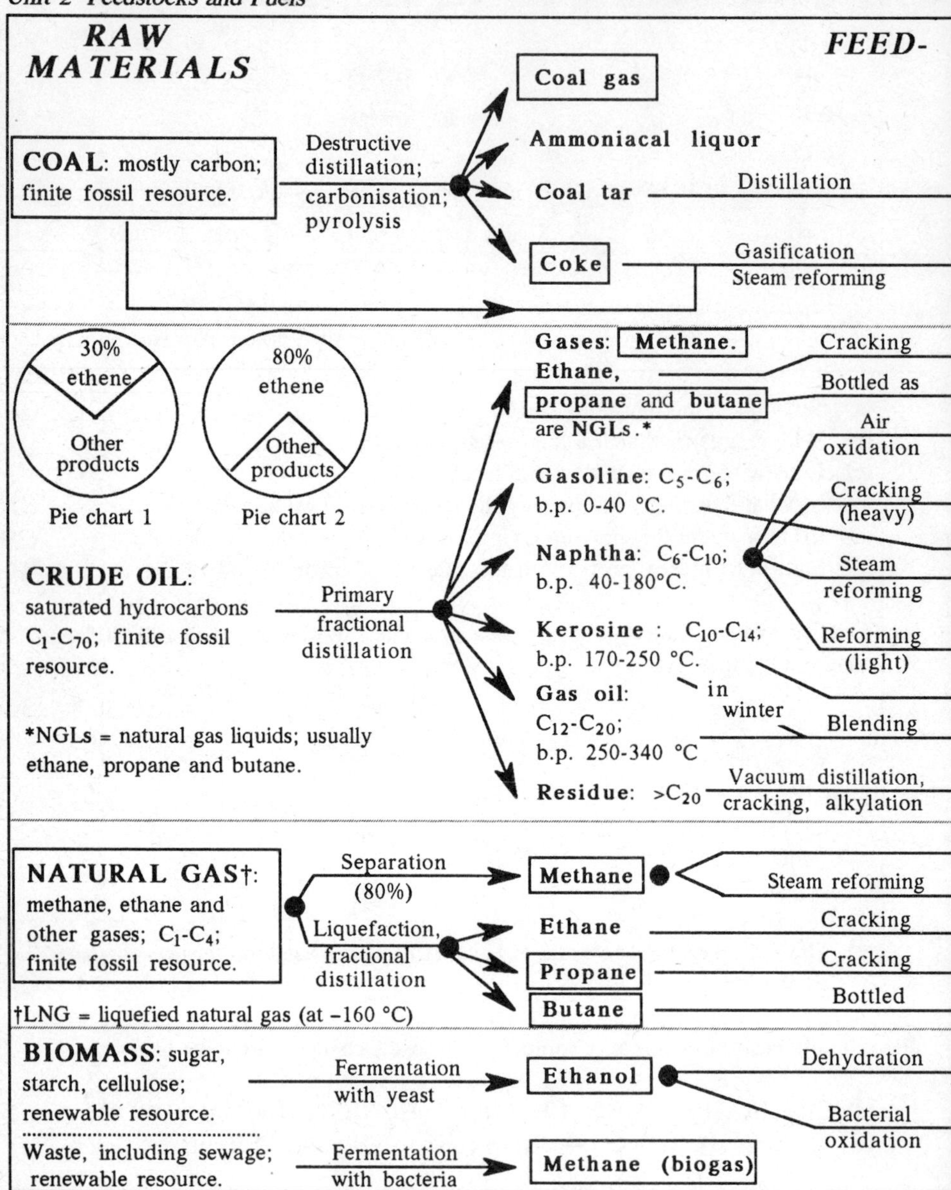
RAW MATERIALS
FEED-
COAL: mostly carbon; finite fossil resource.
Destructive distillation; carbonisation; pyrolysis
Coal gas
Ammoniacal liquor
Coal tar
Distillation
Coke
Gasification
Steam reforming
30% ethene
Other products
Pie chart 1
80% ethene
Other products
Pie chart 2
CRUDE OIL: saturated hydrocarbons C_1-C_{70}; finite fossil resource.
Primary fractional distillation
*NGLs = natural gas liquids; usually ethane, propane and butane.
Gases: Methane.
Ethane,
propane and butane
are NGLs.*
Cracking
Bottled as
Gasoline: C_5-C_6; b.p. 0-40 °C.
Air oxidation
Cracking (heavy)
Naphtha: C_6-C_{10}; b.p. 40-180°C.
Steam reforming
Kerosine : C_{10}-C_{14}; b.p. 170-250 °C.
Reforming (light)
Gas oil: C_{12}-C_{20}; b.p. 250-340 °C
in winter
Blending
Residue: $>C_{20}$
Vacuum distillation, cracking, alkylation
NATURAL GAS†: methane, ethane and other gases; C_1-C_4; finite fossil resource.
Separation (80%)
Liquefaction, fractional distillation
Methane
Steam reforming
Ethane
Cracking
Propane
Cracking
Butane
Bottled
†LNG = liquefied natural gas (at −160 °C)
BIOMASS: sugar, starch, cellulose; renewable resource.
Fermentation with yeast
Ethanol
Dehydration
Bacterial oxidation
Waste, including sewage; renewable resource.
Fermentation with bacteria
Methane (biogas)

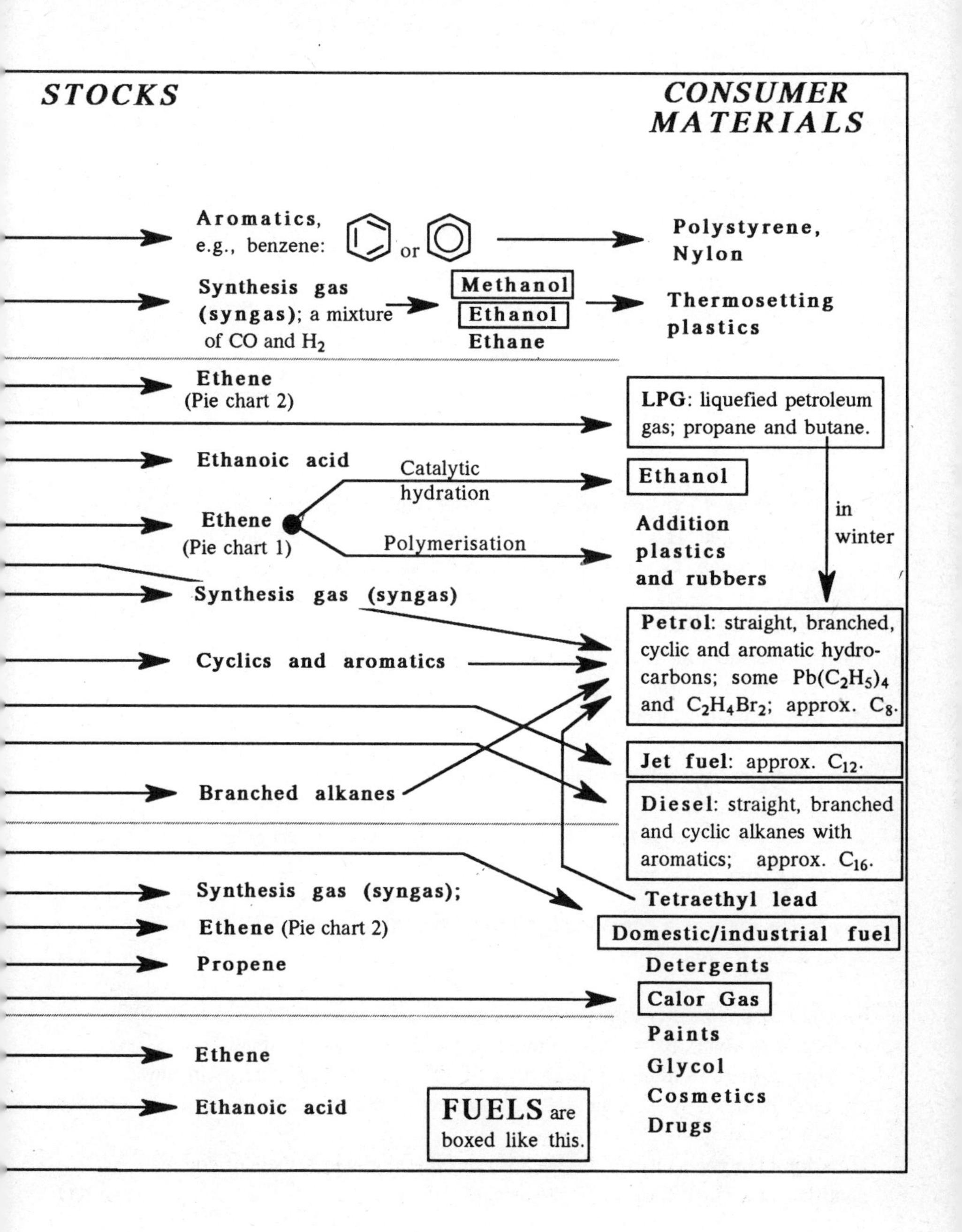
STOCKS
CONSUMER MATERIALS
Aromatics, e.g., benzene: or
Polystyrene, Nylon
Synthesis gas (syngas); a mixture of CO and H_2
Methanol
Ethanol
Ethane
Thermosetting plastics
Ethene (Pie chart 2)
LPG: liquefied petroleum gas; propane and butane.
Ethanoic acid
Catalytic hydration
Ethanol
in winter
Ethene (Pie chart 1)
Polymerisation
Addition plastics and rubbers
Synthesis gas (syngas)
Cyclics and aromatics
Petrol: straight, branched, cyclic and aromatic hydrocarbons; some $Pb(C_2H_5)_4$ and $C_2H_4Br_2$; approx. C_8.
Jet fuel: approx. C_{12}.
Branched alkanes
Diesel: straight, branched and cyclic alkanes with aromatics; approx. C_{16}.
Synthesis gas (syngas);
Tetraethyl lead
Ethene (Pie chart 2)
Domestic/industrial fuel
Propene
Detergents
Calor Gas
Paints
Ethene
Glycol
Cosmetics
Ethanoic acid
FUELS are boxed like this.
Drugs

UNIT 3
MOLECULES TO MOLES

1. Complete the table from (a) to (h).

Quantity	Symbol for quantity	Name of unit	Symbol for unit	Nature of quantity
Mass	m	(a)	kg	Two objects equally difficult to move have the same mass.
Volume	(b)	cubic metre	(c)	(d)
Amount	n	mole	(e)	Two objects composed of the same number of particles are equal in amount.
(f)	Q	(g)	C	(h)

(4)

2. Find the mass of 1 mole of each of the following.

(a) (i) Carbon, C; (ii) copper, Cu; (iii) gold, Au; (iv) monatomic hydrogen, H; (v) hydrogen, H_2; (vi) helium, He; (vii) monatomic oxygen, O; (viii) oxygen, O_2; (ix) ozone, O_3; (x) bromine, Br_2. (10)

(b) (i) Water, H_2O; (ii) pure sulphuric acid, H_2SO_4; (iii) sodium chloride, NaCl; (iv) ethanol, C_2H_5OH. (4)

(c) (i) Sodium ion, Na^+; (ii) chloride ion, Cl^-; (iii) ammonium ion, NH_4^+; (iv) sulphate ion, SO_4^{2-}. (4)

(d) Electricity (electrons), e^-. (1)

3. Name the elements whose molecules are diatomic at r.t.s.p. (7)

4. Calculate the molar mass of (i) cane sugar (sucrose), $C_{12}H_{22}O_{11}$; (ii) haemoglobin, $(C_{759}H_{1208}N_{210}S_2O_{204}Fe)_4$. (2)

5. The molar mass of a compound of phosphorus and hydrogen is 34 g mol^{-1}. What is the formula of this compound? (2)

6. The universe is mostly made of hydrogen. The number of atoms in a gram of hydrogen is about 10^{24}. The number of grams in a star is about 10^{33}. The number of stars in a galaxy is about 10^{11}. The number of galaxies in the universe is about 10^{10}. Calculate the number of atoms in the universe. (1)

7. How long (in years) would it take to count the atoms in a mole of copper, counting at a rate of one atom per second? (1)

8. For carbon dioxide, CO_2, write (i) the molecular mass, (ii) the relative molecular mass, (iii) the mass of one mole, (iv) the molar mass. (2)

9. Calculate the mass of each of the following amounts of substance.

(a) (i) 1 mol magnesium, Mg; (ii) 5 mol monatomic sulphur, S; (iii) 0.10 mol phosphorus, P_4; (iv) 3×10^{-5} mol aluminium, Al. (4)
(b) (i) 0.5 mol copper ion, Cu^{2+}; (ii) 2 mol phosphate ion, PO_4^{3-}. (3)
(c) (i) 10 mmol calcium, Ca; (ii) 5 kmol bromine, Br_2. (2)

10. Calculate the amount of each of the following masses of substance.

(a) (i) 22 g carbon dioxide, CO_2; (ii) 36.5 g hydrogen chloride, HCl; (iii) 80 g argon, Ar; (iv) 120 g iron oxide, Fe_2O_3. (6)
(b) (i) 20 g bromide ion, Br^-; (ii) 0.002 g sulphate ion, SO_4^{2-}. (4)
(c) (i) 100 kg chalk, $CaCO_3$; (ii) 188 mg copper nitrate, $Cu(NO_3)_2$. (4)

11. 'Basic white lead' contains lead carbonate, $PbCO_3$, and lead hydroxide, $Pb(OH)_2$. An experiment was carried out to determine the mole ratio of $PbCO_3$:$Pb(OH)_2$.

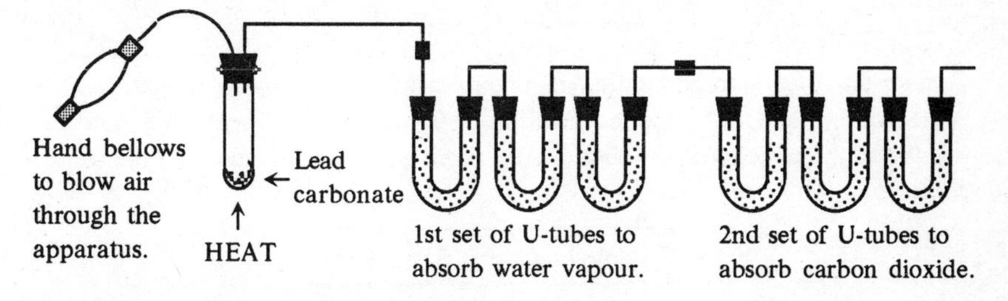

A quantity of basic lead carbonate was heated as shown. Decomposition occurred according to the equations:

$$PbCO_3(s) \longrightarrow PbO(s) + CO_2(g)$$
$$Pb(OH)_2(s) \longrightarrow PbO(s) + H_2O(l).$$

The first set of U-tubes gained 0.27 g, the second set, 1.32 g.

(a) Why is air blown through the apparatus? (1)
(b) Name a solid that could be used in (i) the first set, (ii) the second set of absorption tubes. (2)
(c) (i) Convert 0.27 g water to an amount (in mol). (2)
(ii) Convert 1.32 g carbon dioxide to an amount (in mol). (2)
(iii) State the mole ratio of CO_2:H_2O. (1)
(iv) State the mole ratio of $PbCO_3$:$Pb(OH)_2$. (1)
(v) Write a formula for basic lead carbonate. (1)

12. The molar masses, M_m, of a few hydrates are given in the table.
   Solve for *w, x, y* and *z*.

Formula of hydrate	M_m/g mol^{-1}
$Na_2CO_3.wH_2O$	286
$CuSO_4.xH_2O$	250
$Al(NO_3)_3.yH_2O$	375
$CaSO_4.zH_2O$	145

(8)

13. There are different concentration terms. The most important one in chemistry is *amount of solute per volume of solution*, usually expressed in mol L^{-1}.
   In general, $c = n/V$. Use this relation or other methods to calculate the unknown quantities in the table.

Amount (n)	Volume (V)	Concentration (c) /mol L^{-1}
(a)	0.5 L	4
0.1 mol	250 mL	(b)
20 mol	(c)	5
0.001 mol	0.01 L	(d)
(e)	750 mL	4
0.54 mol	(f)	0.09

(12)

14. A solution in its standard state has a concentration of 1 mol L^{-1}. Let S represent 1 mol L^{-1}.* A concentration of 0.5 mol L^{-1} is then 0.5 S, and so on. State the concentration obtained by mixing solutions 1 and 2 of the same solute.

	Solution 1	Solution 2
(a)	1 L of S	2 L of 2 S
(b)	0.5 L of 2 S	1.5 L of S
(c)	0.25 L of 4 S	0.75 L of S
(d)	100 mL of 10 S	100 mL of S

(4)

**A solution whose concentration is 1 mol L^{-1} is also called a 'molar' solution. For solutions, 'molar' = '1 mol L^{-1}'. The word is also used in molar mass and molar volume (of solids, liquids and gases). But here it has a different meaning: 'molar' now means 'per mole'*

15. When a solution is diluted, its volume goes up and its concentration goes down. The product Vc is constant, i.e., $V_1c_1 = V_2c_2$ where subscripts 1 and 2 refer to before and after dilution (or vice versa). Use this relation or other methods to find the concentration after dilution in each of the following problems.

(Contd.)

(a) 1 L of a 1 mol L^{-1} solution is diluted to 2 L. (1)
(b) 0.5 L of a 2 mol L^{-1} solution is diluted to 4 L. (1)
(c) 1.3 L of a 0.7 mol L^{-1} solution is diluted to 10 L. (2)
(d) 250 mL of a 0.3 mol L^{-1} solution is diluted to 600 mL. (2)
(e) 100 mL of water is added to 600 mL of 0.2 mol L^{-1} solution. (2)

16. The definition of the mole refers to 'elementary entities'. What are the elementary entities of (i) copper, Cu, (ii) neon, Ne, (iii) sodium chloride, NaCl, (iv) water, H_2O, (v) bromide ion, Br^-, (vi) electricity, e^-? (3)

17. (a) Calculate the number of atoms in 3.2 g of oxygen gas. (2)
(b) Calculate the number of ammonium ions in 13.2 g ammonium sulphate. (3)
(c) If a tablespoonful of water (1 mole) were spread uniformly over the surface of our planet, how many water molecules would it give per m^2?
Radius of earth = 6400 km. Surface area of a sphere = $4\pi r^2$. (3)
(d) Calculate the number of electrons in 56 g of (i) iron, (ii) iron(II) ion. (2)
(e) How many electrons are needed to discharge 1 mole of silver ion? (1)
(f) Calculate the mass of potassium nitrate that gives 3 x 10^{23} molecules of oxygen according to the decomposition: $2KNO_3 \rightarrow 2KNO_2 + O_2$. (2)
(g) Calculate the mass of aluminium sulphate that contains 1.8 x 10^{23} sulphate ions. (3)
(h) For 11 g of carbon dioxide, calculate (i) amount, (ii) number of molecules, (iii) number of atoms. (3)
(i) Calculate the amount of calcium chloride that consists of 9 x 10^{23} ions. (2)
(j) Calculate the number of atoms in 1 g of ^{12}C. (1)
(k) Calculate the mass in g of 1 molecule of hydrogen. (2)
(l) Calculate the energy released by the fission of one mole of ^{238}U given that one fission releases 3.04 x 10^{-11} J. (1)
(m) The energy released by the formation of a mole of H–H bond is 435 kJ. What is the energy of one bond only? (1)

18. Vanillaldehyde, $C_8H_8O_3$, is the substance most easily detected by smell: as little as 2 x 10^{-11} g per litre of air can be smelled. Calculate the number of molecules of vanillaldehyde per litre of air at this concentration. (3)

19. X-ray analysis gives the volume of an Na^+Cl^- ion-pair as 4.49 x 10^{-2} nm^3. The density of sodium chloride is 2.17 g cm^{-3}.

(a) What is the volume in cm^3 of one Na^+Cl^- ion-pair? (1)
(b) Calculate the volume in cm^3 of 1 mol of NaCl. (2)
(c) Use your answers to (a) and (b) to calculate the Avogadro number, N_A. (2)

20. Use your data book to find the density and relative atomic mass of copper. Hence calculate the number of atoms in a block of copper measuring 5 cm by 4 cm by 3 cm. (4)

21. (a) What amount of diatomic oxygen, O_2, is obtained from 1 mole of monatomic oxygen, O? (1)
    (b) What mass of diatomic oxygen is obtained from 1 mole of monatomic oxygen? (1)
    (c) What mass of diatomic oxygen is obtained from 8 g monatomic oxygen? (1)

22. (a) What is the amount ratio* of 14 g CO to 44 g CO_2? (1)
    (b) Name a gas that has the same molar mass as carbon monoxide. (1)
    (c) How many Avogadro numbers of ions are there in 5 mol calcium chloride? (1)
    (d) How many water molecules are there in a cup of water (180 g)? (1)
    (e) What mass of nitrogen, N_2, contains the same number of molecules as 16 g oxygen, O_2? (1)
    (f) Calculate the number of sodium ions in ½ litre of 0.5 mol L^{-1} sodium hydroxide solution. (2)
    (g) Calculate the number of atoms in 0.000 059 g nickel. (1)
    *Usually called mole ratio (not molar ratio!).

23. Over a period of time, an unstable element ejected 1.7×10^{17} α particles, each of which becomes a helium atom. 0.0068 cm^3 of helium was collected over the same period. Calculate the Avogadro constant, L. (Take the molar volume of helium as 24 L mol^{-1}.) (3)

24. The molar volume, V_m, of any gas at s.t.p. is 22.4 L mol^{-1}. S.t.p. stands for standard temperature and pressure: 273 K and 1 atm (101 325 Pa). But scientists have decided to change to r.t.s.p.. R.t.s.p. stands for reference temperature and standard pressure: 298 K and 1 bar (100 000 Pa).

    A group of pupils were curious to see how V_m for a gas at r.t.s.p. compared with the value at s.t.p.. They started by weighing known volumes of three gases at r.t.s.p.. Data are shown below.

Gas	Volume /cm^3	Mass /g	M_m /g mol^{-1}
Methane, CH_4	136.5	0.0880	16
Nitrogen, N_2	384.5	0.4341	28
Propane, C_3H_8	226.0	0.4011	44

    (a) Calculate V_m for each gas at r.t.s.p.. (6)
    (b) How does V_m at r.t.s.p. compare with V_m at s.t.p.? (1)
    (c) Give two reasons to explain the answer to (b). (2)
    (d) Plot V_m against T, ignoring the pressure variable. Use the graph to find an approximate value for V_m at room temperature, 20 °C. (3)
    (e) What assumption is made when plotting the the graph in (d)? (1)

25. ***The molar volume of graphite.***

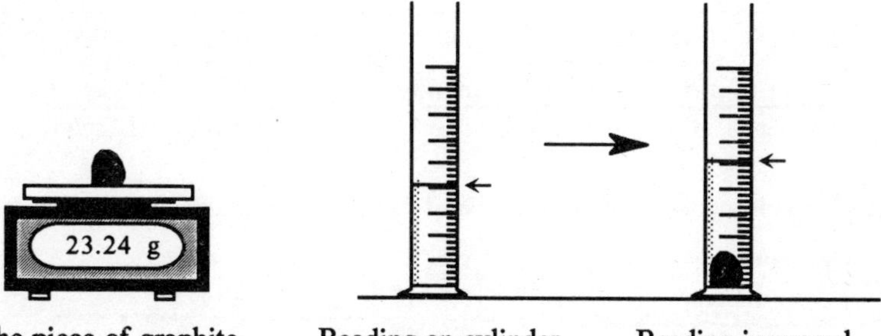

The piece of graphite was weighed. | Reading on cylinder was 40.25 cm^3. | Reading increased to 50.55 cm^3.

A piece of graphite was weighed. Its volume was then found by immersing it in water in a cylinder. Calculate the molar volume of graphite. (3)

26. (a) The V_m of nitrogen (at r.t.s.p.) is 24.8 $dm^3\ mol^{-1}$. Calculate the density of nitrogen in $g\ L^{-1}$. (Note: $dm^3 = 1000\ cm^3 = 1000\ mL = L$.) (2)
(b) The density of hydrogen (at r.t.s.p.) is 0.081 $g\ dm^{-3}$. Calculate the V_m of hydrogen in $dm^3\ mol^{-1}$. (2)
(c) Calculate the density of chloromethane, CH_3Cl, in $g\ L^{-1}$ given the molar volume of the gas (at r.t.s.p.) as 24.8 $L\ mol^{-1}$. (2)

27. The measuring cylinders below contain one mole each of the first four members of the homologous series of alcohols.

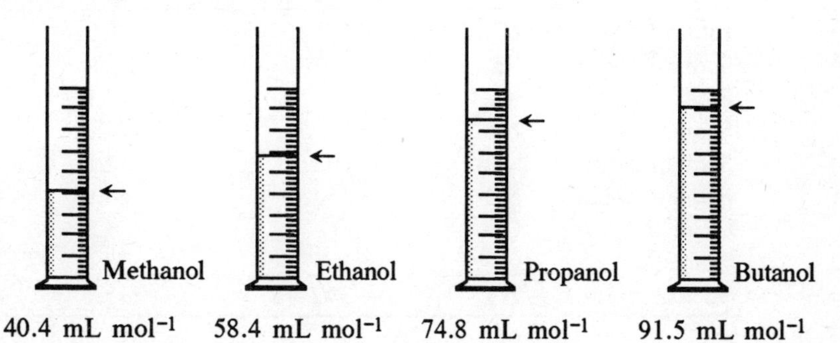

40.4 $mL\ mol^{-1}$ 58.4 $mL\ mol^{-1}$ 74.8 $mL\ mol^{-1}$ 91.5 $mL\ mol^{-1}$

(a) Explain the (almost) regular increase in molar volume from methanol to butanol. (2)
(b) Calculate the average molar volume of the 'methene' group, $-CH_2-$, in $mL\ mol^{-1}$. (2)
(c) Estimate the volume of one mole of hexanol. (1)

28. Use Avogadro's law to find the values of ratios (a), (b), (c) and (d).

(a) $\dfrac{\text{Number of molecules in 3 L hydrogen}}{\text{Number of molecules in 4 L oxygen}}$ (b) $\dfrac{\text{Mass of 1 L nitrogen}}{\text{Mass of 1 L carbon dioxide}}$

(c) $\dfrac{\text{Number of atoms in 2 L of methane}}{\text{Number of atoms in 1 L of ethane}}$ (d) $\dfrac{\text{Mass of 2 L helium}}{\text{Mass of 1 L neon}}$ (4)

29. The molar volume of nitrogen was found using the apparatus shown below.

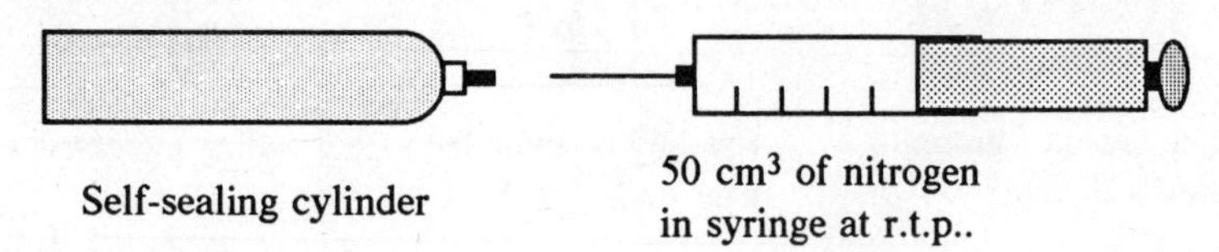

Self-sealing cylinder

50 cm³ of nitrogen in syringe at r.t.p..

50 cm³ of nitrogen at r.t.p. were injected into a weighed self-sealing cylinder, which was then reweighed.

Results			
	Mass of empty cylinder	=	10.3942 g
	Mass of cylinder plus nitrogen	=	10.4538 g

Calculate the volume of one mole of the gas at r.t.p. (3)

30.

	Solid	Formula	Molar mass, M_m /g mol^{-1}	Molar volume, V_m /cm³ mol^{-1}
(a)	Lithium	Li	7	13.1
	Sodium	Na	23	23.7
	Potassium	K	39	44.9
(b)	Sodium chloride	NaCl	58.5	27.0
	Sodium bromide	NaBr	103	32.1
	Sodium iodide	NaI	150	40.9
(c)	Graphite	C	12	5.3
	Diamond	C	12	3.4
	Sucrose	$C_{12}H_{22}O_{11}$	342	215.1

Explain the variation in molar volumes within (a), within (b) and within (c). (3)

(d) Why is the volume of one mole of sucrose much greater than that of any other substance in the table? (1)

(e) Calculate the densities of graphite and diamond from the data in the table and explain how these confirm your answer to (c). (4)

31. (a) What volume of SO_2 has the same mass as 24.8 L of O_2 at r.t.s.p.? (V_m at r.t.s.p. is 24.8 L mol^{-1}.) (1)
    (b) What mass of CO_2 has the same volume as 8 g of O_2 at r.t.s.p.? (1)

32. Calculate the volume of oxygen evolved by the decomposition of 1 L of a 1 mol L^{-1} solution of hydrogen peroxide, H_2O_2. Take $V_m(O_2)$ as 24.0 L mol^{-1}. (2)

33. Hydrogen and helium, H_2 and He, are both light gases used in balloons, but hydrogen is half as dense as helium. Hydrogen-filled balloons, therefore, can lift heavier loads. The problem, however, is "how much heavier?" Twice? This seems plausible. But plausible arguments are often wrong. It is wiser to work out the answer.

    To do so, we make three assumptions: first, that the fabric of the balloon is so thin that it has no mass; second, that V_m of a gas is 25 L mol^{-1}; third, that air is 100% N_2. (It is in fact approx. 80% N_2 and 20% O_2.)

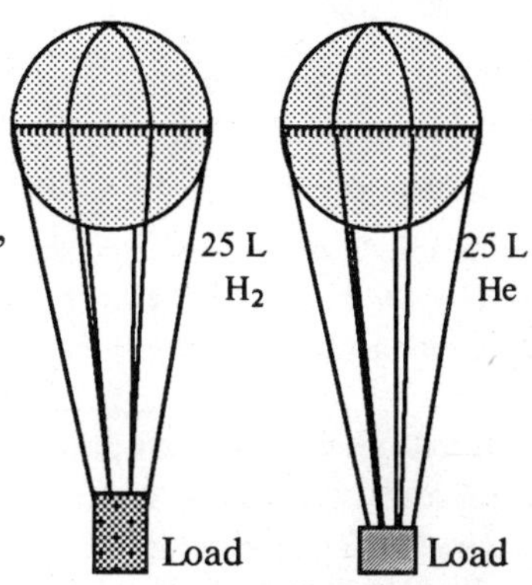

Can a hydrogen-filled balloon lift twice as much as a helium-filled balloon?

    (a) What is the mass of
        (i) 25 L of hydrogen; (1)
        (ii) 25 L of helium; (1)
        (iii) 25 L of air (as N_2)? (1)
    (b) When the balloon is filled with hydrogen,
        (i) how much lighter is it than when filled with air (as N_2); (1)
        (ii) what load would just stop it rising? (1)
    (c) When filled with helium,
        (i) how much lighter is it than when filled with air (as N_2); (1)
        (ii) What load would just stop it rising? (1)
    (d) Compare your answers to (b)(ii) and (c)(ii). Can the hydrogen-filled balloon lift twice as much as a helium-filled balloon? If not, how much more? (2)
    (e) (i) This has been a 'theoretical' proof. Explain. (1)
        (ii) By what other method might the problem be resolved? (1)

34. The charge (Q) is the product of electric current (I) and time (t): $Q = It$.

    (a) Calculate the charge passing a point in a circuit when 4 A flow for 100 s. (1)
    (b) Calculate the current in a circuit when a charge of 60 C is delivered in 10 s. (1)
    (c) How long does it take to deliver 33 C using a current of 3 A? (1)

35. Calculate the volume of a mole of water in cm^3 allowing 0.016 nm^3 for the volume of an oxygen atom and 0.007 nm^3 for that of a hydrogen atom. (3)

36. Calculate the number of singly-charged positive ions arriving per second at the detector in a mass spectrometer when the current registered is 10^{-10} A. (2)

37. Calculate the current needed to deposit 1.19 g of tin from molten tin(II) chloride in 20 minutes. (3)

38. Faraday's first law states that during electrolysis *the mass of a metal deposited is proportional to the charge used to deposit the metal.* Why would replacing '*mass*' by '*amount*' be correct? (1)

39. Faraday's second law states that during electrolysis *one mole of a metal is deposited by 96 500 C or by a simple multiple of 96 500 C.* Explain why '*or by a simple multiple of 96 500 C*' must be included in the statement. (2)

40. An electric current was passed through two cells, A and B, in series. Cell A contains aqueous silver ion. Cell B contains acidified copper ion.
    0.032 g of silver was deposited on the cathode of cell A.

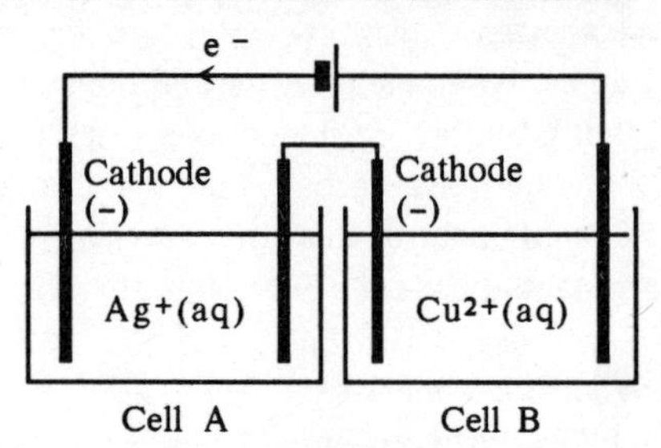

(a) Write ion-electron equations for the cathode half-reactions. (2)
(b) (i) What mass of copper should be deposited in cell B? (3)
    (ii) The mass of copper deposited was 0.0091 g. Calculate the % yield. (1)
(c) The 0.032 g of silver required a current of 40 mA for 11 min 55 s. Calculate the Avogadro constant, L, (given the charge, e, of an electron as 1.6×10^{-19} C and the Faraday constant, F, as 96 500 C mol^{-1}). (3)

41. The electrode half-reactions during the electrolysis of dilute sulphuric acid are

$$2H^+(aq) + 2e^- \rightarrow H_2(g)$$

and

$$2OH^-(aq) \rightarrow H_2O(l) + {}^1\!/_2O_2(g) + 2e^-.$$

(a) Calculate the volume of hydrogen liberated by
    (i) N_A e^-; (2)
    (ii) 30 mA flowing for 30 min. (3)
(b) Calculate the volume of oxygen liberated by
    (i) Q = 96 500 C; (2)
    (ii) I = 0.01 A flowing for t = 3000 s. (3)

Take V_m(any gas) as 24.8 L mol^{-1}.

42. A current of 5 A was passed through molten sodium chloride for 30 min.

 (a) Give ion-electron equations for the electrode half-reactions. (2)
 (b) Calculate
  (i) the charge used; (1)
  (ii) the mass of sodium liberated; (2)
  (iii) the volume of chlorine evolved at r.t.s.p. (V_m = 24.8 dm^3 mol^{-1}.) (2)

43. ***Electrolytic refining of copper***
Thin starting sheets of pure copper are lowered into an electrolyte. The sheets each weigh 2 kg. Thick sheets of impure (blister) copper are inserted between the starting sheets.

Electrolysis transfers copper from impure sheets to starting sheets. After 2 weeks, the starting sheets each weigh 100 kg.

Electrolytic copper is very pure because impurities in the blister copper fall to the floor of the cell as an insoluble deposit called anode mud. This mud is removed and worked for precious metals.

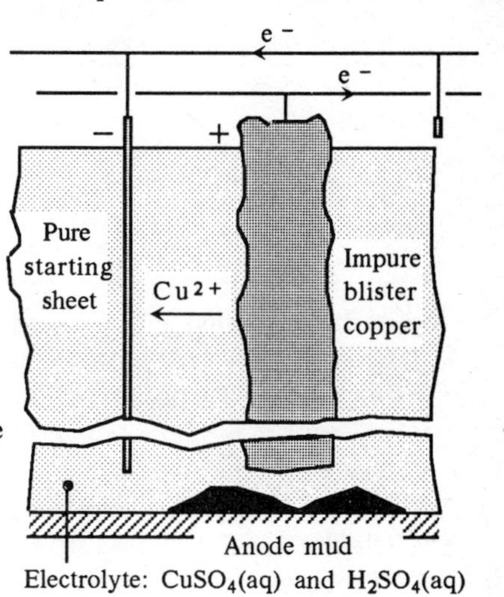

Electrolyte: $CuSO_4$(aq) and H_2SO_4(aq)

 (a) Write ion-electron equations for the half reactions at
  (i) the negative electrode (cathode); (1)
  (ii) the positive electrode (anode). (1)
 (b) Calculate the steady current used over the two week period. (3)
 (c) (i) Name two precious metals to be found in the anode mud. (1)
  (ii) Offer a reason why these metals are found with copper. (1)
 (d) Over a period of time, 2.3 tonnes of blister copper gave 2.1 tonnes of refined copper. Calculate the purity of the blister copper. (1)

44. Electrolysis of a molten salt with a current of 1 A liberated 0.97 g of the metal in 39.92 min. Calculate the r.a.m. (A_r) of the metal and identify it. (4)

45. Aluminium is deliberately corroded to improve its corrosion resistance! The process is called anodising: the oxide layer on the metal's surface is thickened by making it the anode in an electrolytic cell:

$$2Al(s) + 3H_2O(l) \rightarrow Al_2O_3(s) + 6H^+(aq) + 6e^-$$

State (i) the charge used to make 1 mol of aluminium oxide; (1)
 (ii) the amount of hydrogen ion created by this charge. (1)

46. ***Aluminium smelting in the West Highlands***
The Lochaber cell rooms contain a line of 80 cells connected in series. These operate with a current of 175 000 A. Calculate the total mass of aluminium produced per day. (4)

47. The table gives information on the combustion of gases in excess oxygen.

Gas burned			Volume of O_2/mL
Name	Formula	Volume/mL	
(a) Hydrogen	H_2	10	18
(b) Carbon monoxide	CO	20	37
(c) Methane	CH_4	30	74
(d) Ethene	C_2H_4	40	210
(e) Propyne	C_3H_4	50	330
(f) Ammonia	NH_3	60	145

For each combustion
(i) write a balanced equation, including state symbols; (6)
(ii) state the composition of the product at 99 °C; (12)
(iii) state the composition of the product at 101 °C. (6)

48. (a) Write a balanced equation for the complete combustion of butane, C_4H_{10}. (1)
(b) Use the equation to determine the composition of the gas mixture obtained by burning 10 cm^3 C_4H_{10} in 90 cm^3 O_2 (i) at 99 °C and (ii) at 101 °C. (2)

49. 5 g of magnesium are added to 100 mL of 2 mol L^{-1} HCl.

(a) (i) Write a traditional equation for the reaction; (1)
(ii) Write an ionic equation for the reaction. (1)
(b) Which reactant is in excess? Show your working. (3)
(c) What volume of hydrogen is evolved at r.t.s.p.? (V_m = 24.8 L mol^{-1}.) (2)

50. 11 540 mL of O_2 are liberated at r.t. by the decomposition of 1 L of 1 mol L^{-1} H_2O_2. Calculate V_m of O_2 at r.t.. (3)

51. For 2 L of ethane, calculate the volume of oxygen needed for
(i) partial combustion (to CO and H_2O); (3)
(ii) complete combustion (to CO_2 and H_2O). (3)

52. 10 mL of a gaseous hydrocarbon were burned in excess oxygen. 40 mL of carbon dioxide and 40 mL of steam were obtained. Determine the molecular formula of the hydrocarbon. (3)

53. Nitrogen monoxide reacts on contact with oxygen to give nitrogen dioxide, an acidic gas that is very soluble in water.

$$NO(g) + \tfrac{1}{2}O_2(g) \rightarrow NO_2(g).$$

100 mL NO were mixed with 80 mL O_2. The product was passed through water.

(a) State the volume of the gas mixture after reaction. (1)
(b) What is the final volume after passing through water? (1)

54. 9 mL of a gaseous hydrocarbon on combustion with 50 mL oxygen (in excess) gave 32 mL of gaseous product at r.t.. The volume of this product decreased to 5 mL when potassium hydroxide (which absorbs CO_2) was introduced. What is the molecular formula of the hydrocarbon? (4)

55. Diborane (B_2H_6) is a gas that burns in oxygen to give boron oxide, which is a solid, and water. A mixture of 20 cm^3 diborane and 100 cm^3 oxygen was ignited in a eudiometer (an apparatus used in gas analysis).

(a) Write a balanced equation (with state symbols) for the reaction. (1)
(b) Determine the composition of the products at 298 K. (2)

56. Benzene is a valuable feedstock. It is made by reforming naphtha:

$$C_6H_{14}(g) \longrightarrow C_6H_6(g) + 4H_2(g) \quad \ldots 1$$

Toluene, for which there is less demand, is produced at the same time,

$$C_7H_{16}(g) \longrightarrow C_6H_5CH_3(g) + 4H_2(g) \quad \ldots 2$$

but is converted to benzene by dealkylation by heating with hydrogen:

$$C_6H_5CH_3(g) + H_2(g) \longrightarrow C_6H_6(g) + CH_4(g) \quad \ldots 3$$

(a) State the volume ratio of reactants:products for reactions 1, 2 and 3. (3)
(b) Dealkylation of 1 tonne of toluene gives 0.77 tonne of benzene. Calculate the percentage yield. (3)

57. Construct ion-electron equations for the half-reactions shown below and say whether they are oxidations or reductions.

(a) $BiO^+ \rightarrow Bi$
(b) $HOBr \rightarrow Br_2$
(c) $BrO_3^- \rightarrow \tfrac{1}{2}Br_2$
(d) $FeO_4^{2-} \rightarrow Fe^{3+}$
(e) $NO \rightarrow HNO_2$
(f) $PH_3 \rightarrow P$
(g) $H_3PO_3 \rightarrow H_3PO_4$
(h) $Pb^{2+} \rightarrow PbO_2$
(i) $SO_4^{2-} \rightarrow SO_3^{2-}$
(j) $S_2O_3^{2-} \rightarrow S_4O_6^{2-}$ (10)

58. Write an ion-electron equation for

(a) the oxidation of any Group 2 element to its oxide. (1)
(b) the reduction of any Group 6 element to its hydride. (1)
(c) the oxidation of H_2O_2 to oxygen. (1)
(d) the reduction of H_2O_2 to hydrogen. (1)
(e) the reduction of H_2O_2 to water. (1)

59. Ion-electron equations for seven half-reactions are shown below.

1. $Zn^{2+} + 2e^- \rightleftharpoons Zn$
2. $2H^+ + 2e^- \rightleftharpoons H_2$
3. $Cu^{2+} + 2e^- \rightleftharpoons Cu$
4. $Fe^{3+} + e^- \rightleftharpoons Fe^{2+}$
5. $Br_2 + 2e^- \rightleftharpoons 2Br^-$
6. $Cl_2 + 2e^- \rightleftharpoons 2Cl^-$
7. $MnO_4^- + 8H^+ + 5e^- \rightleftharpoons Mn^{2+} + 4H_2O$

The forces with which forward reactions occur increase down ↓ the list.

Construct balanced ionic equations from (a) 1 and 2, (b) 1 and 3, (c) 1 and 4, (d) 1 and 5, (e) 2 and 6, (f) 4 and 6, (g) 4 and 7. (14)

60. In titrimetric analysis a reagent A in a burette is titrated against a known volume of a reagent B in a flask.* The concentration of A or B, whichever is unknown, is worked out from the reading (titre) on the burette. The relation

$$\frac{V_1 c_1}{n_1} = \frac{V_2 c_2}{n_2}$$

is useful for working out the unknown concentration. Subscripts 1 and 2 refer to the two reagents (reactants) involved in the titration (reaction).

The table below gives data on four separate titrations. Calculate the unknown concentration in each titration.

Titration	Reagent A Titre V_1 /mL	Reagent A c_1 /mol L⁻¹	Reagent B V_2 /mL	Reagent B c_2 /mol L⁻¹	Equation	n_1 /mol	n_2 /mol
(a)	16.8	0.1	20	c	A + 2B → Products	1	2
(b)	22.4	0.15	15	c	2A + B → Products	2	1
(c)	12.7	c	25	0.25	A + 5B → Products	1	5
(d)	24.6	c	20	0.06	2A + 3B → Products	2	3

*A reagent is a solution of a reactant. (12)

61. *Determination of x in $FeSO_4.xH_2O$*

Approximately 6 g of the hydrated iron(II) sulphate were weighed off accurately by difference to give 6.069 g. This mass was dissolved in freshly boiled and cooled water and made up to 100 mL.

A burette was filled with a reference solution of potassium permanganate of c = 0.05 mol L^{-1}.

20 mL of the iron(II) solution were pipetted into a conical flask containing dilute sulphuric acid.

The contents of the flask were titrated with the permanganate solution until a faint permanent pink. This is the end-point. The reading on the burette was noted. The titration was repeated twice.

Results

Titration	Volume of MnO_4^-(aq)/mL
First	18.1
Second	17.5
Third	17.4

MnO_4^-(aq)
c = 0.05
mol L^{-1}

Equation

$$MnO_4^- + 8H^+ + 5Fe^{2+} \rightarrow Mn^{2+} + 4H_2O + 5Fe^{3+}$$

Fe^{2+}(aq)

(a) Explain "*Approximately* ... 6 g were weighed off *accurately* ...". (1)

(b) Explain the term "*weighing by difference*". (1)

(c) Why should freshly boiled water be used to dissolve the iron(II) compound? (1)

(d) Why must sulphuric acid be present during this titration? (1)

(e) This titration is 'self-indicating': it does not need an indicator. Explain. (1)

(f) State the volume of permanganate solution to use in the calculation. (1)

(g) (i) Calculate the concentration of the iron(II) solution in mol L^{-1} from the results of the titration. (3)

(ii) State the mass of iron(II) compound in one litre as made up. (1)

(iii) From (i) and (ii) calculate the mass of one mole of iron(II) compound. (1)

(iv) Calculate the mass of one mole of the compound as an expression in x from the formula $FeSO_4.x\,H_2O$. (1)

(v) Use your answers to (iii) and (iv) to solve for x. (1)

(vi) Write the formula for hydrated iron(II) sulphate. (1)

UNIT 4

BIOMOLECULES

1. Lipids are substances, found in living tissue, that are soluble in organic solvents but not water. Different kinds of lipid have little in common with each other apart from solubility. Fats and oils, essential oils (essences), steroids, waxes and some vitamins are lipids.

Structural formulae for four lipids are shown below. Three are triglycerides and one a steroid (cholesterol).

$$\begin{array}{l} CH_2OCO(CH_2)_{16}CH_3 \\ | \\ CHOCO(CH_2)_{16}CH_3 \\ | \\ CH_2OCO(CH_2)_{14}CH_3 \end{array}$$

Triglyceride A

$$\begin{array}{l} CH_2OCO(CH_2)_7CH{=}CH(CH_2)_7CH_3 \\ | \\ CHOCO(CH_2)_{14}CH_3 \\ | \\ CH_2OCO(CH_2)_7CH{=}CH(CH_2)_5CH_3 \end{array}$$

Triglyceride B

$$\begin{array}{l} CH_2OCO(CH_2)_4CH{=}CHCH_2CH{=}CHCH_2CH{=}CH(CH_2)_4CH_3 \\ | \\ CHOCO(CH_2)_{16}CH_3 \\ | \\ CH_2OCO(CH_2)_3CH{=}CHCH_2CH{=}CHCH_2CH{=}CHCH_2CH{=}CHCH_2CH{=}CHCH_2CH_3 \end{array}$$

Triglyceride C

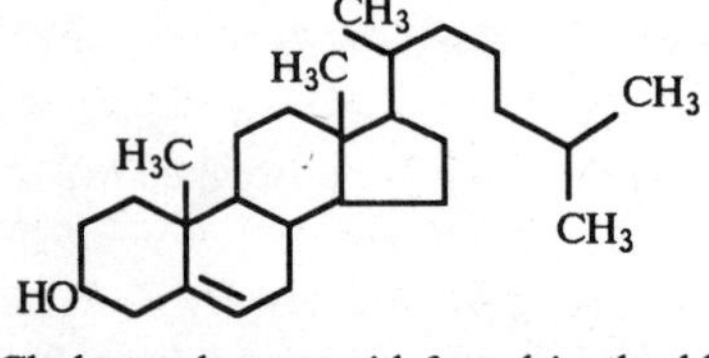

Cholesterol, a steroid found in the blood, brain and most other tissues of the body.

(a) Define the term 'triglyceride'. (1)

(b) Which triglyceride is
(i) saturated, (1)
(ii) polyunsaturated, (1)
(iii) most likely to be found in fish oil? (1)

(c) (i) Identify the triglyceride that would react with no more than two moles of diatomic hydrogen per mole of triglyceride. (1)
(ii) Explain your choice in (i). (1)
(iii) Name the reaction in (i). (1)
(iv) Draw a short structural formula for the product in (i). (1)

(d) (i) Which material in food increases blood cholesterol? (1)
(ii) Too much cholesterol in the blood may cause disease. What is the disease? (1)
(iii) According to medical opinion, how might the risk of this disease be reduced? (1)

(e) (i) Is cholesterol a primary, secondary or tertiary alcohol? (1)
(ii) Write the molecular formula of cholesterol. (1)

2. Fats and oils come from plants and animals. The animals may be land animals or marine animals.

   (a) Classify the following fats and oils, with respect to origin, in a table with suitable headings: lard, palm kernel oil, olive oil, rapeseed oil, tallow, castor oil, cod liver oil, suet, butter, linseed oil, herring oil, halibut oil, corn oil. (3)

   (b) Fats and oils are lipids. Use the classification in (a) to state how the origin of these lipids determines their state at room temperature? (1)

3. 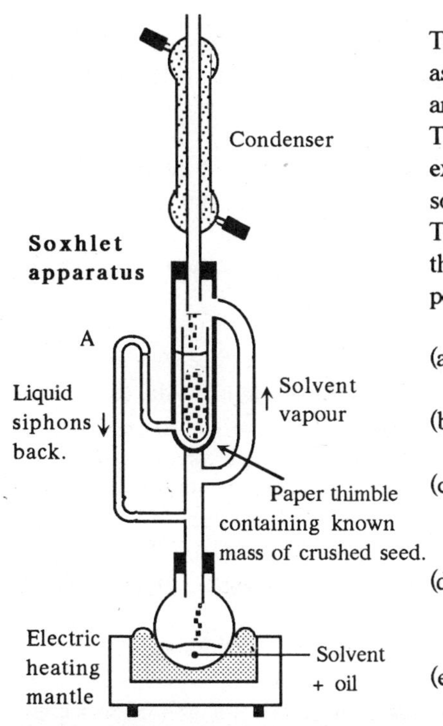

The food industry employs chemists as analysts to check that raw materials and products meet required standards. The oil content of sunflower seed, for example, may be determined by soxhlet extraction, as shown opposite. The main feature of the apparatus is the use of a siphon (A), which periodically returns liquid to the flask.

   (a) Name a solvent that could be used in this extraction. (1)

   (b) When must an electric mantle rather than a bunsen be used for heating? (1)

   (c) Soxhlet extraction has a major advantage over trituration*. Explain. (2)

   (d) Describe a simple test to determine whether extraction is complete or incomplete. (2)

   (e) How would you determine the mass of oil extracted from the seed? (2)

*Trituration: grinding with solvent and pouring off or filtering off the solvent.

4. About fifty fatty acids occur naturally in fats and oils. Two have been identified as <u>precursors</u> of two series of <u>essential fatty acids</u>, EFAs. The first is linoleic acid, octadec-9,12-dienoic acid, and the second, linolenic acid, octadec-9,12,15-trienoic acid. The EFAs are all <u>polyunsaturated acids</u>.

   (a) Explain the underlined words. (3)

   (b) Draw short structural formulae for linoleic and linolenic acid. (2)

5. The short structural formulae and melting points of four acids are shown below.

Name	Short structural formula	M.p./°C
Stearic acid	$CH_3(CH_2)_{16}COOH$	70
Oleic acid	$CH_3(CH_2)_7(CH{=}CHCH_2)(CH_2)_6COOH$	16
Linoleic acid	$CH_3(CH_2)_4(CH{=}CHCH_2)_2(CH_2)_6COOH$	−5
Linolenic acid	$CH_3CH_2(CH{=}CHCH_2)_3(CH_2)_6COOH$	−11

(a) Which variable is responsible for the change in melting points? (1)

(b) (i) Should a plot of this variable (*x* axis) against m.p. (*y* axis) be done as a line graph, spike graph or bar graph? (1)

(ii) Draw the graph. (2)

(iii) Predict the m.p. of the C_{18} acid with four double bonds $molecule^{-1}$. (1)

(c) (i) Explain how difference in molecular shape is important in determining the m.p.s of stearic and oleic acids. (3)

(ii) Elaidic acid has the same short structural formula as oleic acid, but melts at 44 °C instead of 16 °C! Offer a reason for this. (3)

6. CH_2OH
|
$CHOH$
|
CH_2OH
Glycerol

A *simple* triglyceride is one in which the three fatty acids condensed with glycerol are the same. In tristearin, for example, three molecules of stearic acid, $C_{17}H_{35}COOH$, are condensed with one of glycerol; in triolein, three molecules of oleic acid, $C_{17}H_{33}COOH$, are condensed with one of glycerol.

(a) Draw short structural formulae of (i) tristearin, (ii) triolein. (2)

(b) Describe a chemical test to show that tristearin is saturated and triolein unsaturated. (2)

(c) In what way would tristearin and triolein be physically different? (1)

7. Glycerol | Stearic acid residue

Glyceryl monostearate

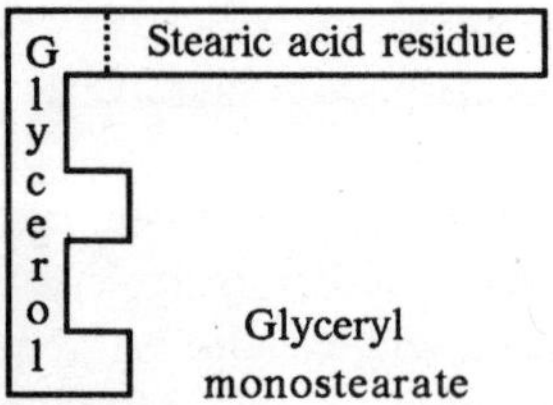

Glyceryl monostearate is a monoglyceride widely used as an emulsifying agent in foods. It is made from fully hardened fat, mostly glyceryl tristearate, by a reaction called 'trans-esterification': one mole of glyceryl tristearate is heated to 200 °C with two moles of glycerol in the presence of a catalyst.

(a) Explain the underlined words. (1,2)

(b) Write an equation for the trans-esterification reaction using the same kind of representation as used above (but omit the labelling). (2)

(c) Construct a flow diagram for the process. (3)

8. Margarines became commercially important in the early 20th Century after the discovery that hydrogenation could change plant oils into fats.

   Hydrogenation: $-CH=CH- \; + \; H_2 \longrightarrow -CH_2CH_2-$

   The fats are suitable for margarine provided hydrogenation is not complete. The progress of hydrogenation can be followed by determining iodine values: *the iodine value (I.V.) of an oil or fat is the mass of iodine in grams that reacts with 100 g of the oil or fat* (quoted as a number only).

   Halogenation: $-CH=CH- \; + \; I_2 \longrightarrow -CHICHI-$

   Iodine values are determined with Wijs reagent, a reactive solution of iodine of known concentration. A measured volume of the reagent is added to a known mass of the oil or fat. After a time the unreacted iodine is determined by titration with sodium thiosulphate solution. The iodine value is then worked out.

   (a) (i) Why was the discovery of hydrogenation important to the margarine industry? (1)
   (ii) Name the catalyst used in the process. (1)
   (iii) Explain why complete hydrogenation of an oil would make it unsuitable for margarine. (1)

   (b) (i) What happens to the I.V. of an oil or fat as hydrogenation progresses? (1)
   (ii) What would be the I.V. of a fully saturated fat? (1)

   0.56 g of fat was dissolved in some tetrachloromethane in a flask. 25 mL of Wijs reagent, equivalent to a 0.1 mol L^{-1} solution of I_2, were added. The flask was left for 1 hr. The unabsorbed iodine was then titrated with 0.1 mol L^{-1} $Na_2S_2O_3$ and required 23.4 mL of this solution.

   $$I_2 + 2Na_2S_2O_3 \rightarrow 2NaI + Na_2S_4O_6$$

   (c) What mass of iodine was added to the fat? (2)

   (d) (i) What mass of sodium thiosulphate is contained in 23.4 mL of solution? (2)
   (ii) Use your answer to (i) and the above equation to calculate the mass of unabsorbed iodine. (2)
   (iii) State the mass of iodine that reacted with 0.56 g of fat. (1)
   (iv) Calculate the I.V. of the fat, giving your answer as the nearest whole number only. (1)

9. Esters contain the group $-C(=O)-O-$.

   (a) Use full structural formulae to write an equation for the reaction between propanoic acid and methanol and circle the ester group in the product. (1)

   (b) Name the ester produced in (a). (1)

   (c) The structural formula of propanoic acid also contains the above group. But this does not make propanoic acid an ester. Explain. (2)

10. Name A, B and C in the following:

(a) Ethanoic acid + propanol → A + water. (1)
(b) Propanoic acid + B → ethyl propanoate + water. (1)
(c) C + butanol → butyl methanoate + water. (1)

11. Draw structural formulae for the esters in (a), (b) and (c) of Question 10, using short formulae for the alkyl groups and full formulae for the ester groups. (3)

12. ^{18}O is a heavier isotope of oxygen. Methanol that contains extra ^{18}O may be written as shown below. Ethanoic acid that contains no extra ^{18}O is written in the usual way.

The ester that forms when these substances condense contains the extra ^{18}O while the water formed at the same time does not.

$$H_3C{-}C(=O){-}O{-}H \qquad H_3C{-}{}^{18}O{-}H$$

Write an equation for the condensation, using this evidence to show where cleavage of the acid and alcohol molecules occurs. (2)

13. Carboxylic acid + Alcohol $\underset{\text{Reverse reaction}}{\overset{\text{Forward reaction}}{\rightleftharpoons}}$ Ester + Water

Name the forward and reverse reactions in the above word equation. (1)

14. The general formula for an ester is XCOOY where X and Y are the same or different groups. Examine the structural formulae of the four esters shown below.

1. Methyl salicylate (oil of wintergreen): $C_6H_4(OH){-}C(=O){-}O{-}CH_3$

2. Tripalmitin, a fat: $CH_2{-}O{-}C(=O){-}C_{15}H_{31}$ / $CH{-}O{-}C(=O){-}C_{15}H_{31}$ / $CH_2{-}O{-}C(=O){-}C_{15}H_{31}$

3. Ethyl methanoate, rum flavour: $H{-}C(=O){-}O{-}C_2H_5$

4. Acetyl salicylic acid, aspirin: $C_6H_4(C(=O){-}O{-}H){-}O{-}C(=O){-}CH_3$

(a) Copy the following table headings and complete the table for each of the esters above.

Ester	Short structural formula of: X	Y	Acid	Alcohol

(16)

(b) Draw the full structural formula of the alcohol in ester 2. (1)
(c) These esters belong to other classes of compound. In each case name the other class of compound. (4)

15.

Ester	Short formula	B.p./°C
Ethyl methanoate	$HCOOC_2H_5$	54
Ethyl ethanoate	$CH_3COOC_2H_5$	77
Ethyl propanoate	$C_2H_5COOC_2H_5$	99
Ethyl butanoate	$C_3H_7COOC_2H_5$	120
Ethyl pentanoate	$C_4H_9COOC_2H_5$	145

The formulae and b.p.s of some esters are shown in the table.

(a) Use these data to predict the b.p. of ethyl hexanoate. (1)
(b) Draw the full structural formula of ethyl butanoate. (1)

16.
1. $C_5H_{11}-O-C(=O)-H$
2. $C_3H_7-C(=O)-O-CH_3$
3. $C_2H_5-O-C(C_4H_9)=O$
4. $O=C(C_5H_{11})-O-C_3H_7$
5. $CH_3COOC_6H_{13}$
6. $C_4H_9OCOC_2H_5$

Short formulae for six esters are shown above. Each gives two products when boiled with sodium hydroxide solution.

(a) Name the reaction that occurs with each ester. (1)
(b) (i) Write short formulae for the products; (6)
(ii) name the products. (6)

17. ***Terylene*** : <u>linear</u> polyester; <u>thermoplastic</u>; <u>synthetic</u>; origin: Great Britain; discovery: 1939-41; <u>monomers</u>: terephthallic acid and glycol; polymerisation: <u>condensation</u>; principal use: fibre in textile industry.

$HO-C(=O)-C_6H_4-C(=O)-OH$ $\qquad$ $HO-CH_2CH_2-OH$

Terephthallic acid
a dicarboxylic acid.

Glycol, a diol,
a dihydric alcohol.

(a) Explain the underlined words. (5)
(b) Give the systematic names of (i) terephthallic acid, (ii) glycol. (2)
(c) (i) Draw a portion of the polyester molecule derived from three monomers. (1)
(ii) Draw a box around one ester group. (1)
(iii) Underline the repeating unit. (1)
(d) Name the raw material from which the feedstocks terephthallic acid and glycol are derived. (1)
(e) J.R. Whinfield's discovery of terylene was based on earlier work by W.H. Carothers (the discoverer of nylon). Carothers had attempted to make a fibre by condensing phthallic acid (an isomer of terephthallic acid) with glycol. But the polyester he obtained did not give a fibre. Explain why terephthallic acid gives a good fibre while phthallic acid does not. (3)

$HO-C(=O) \quad C(=O)-OH$ (on adjacent ring positions)

Phthallic acid

18. ***Polyester resin*** : three-dimensional polymer; thermosetting; synthetic; monomers (various): terephthallic acid, maleic acid and glycol for polyester, styrene for crosslinking; polymerisation: condensation for ester links, addition for cross-links; principal use: boat-hulls and car-bodies with glass fibre reinforcement (glass reinforced plastic, GRP).

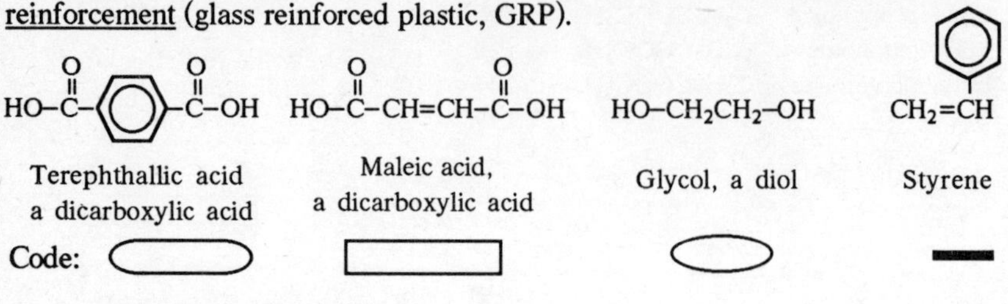

(a) Explain the underlined words. (5)

(b) Give the systematic name of styrene. (1)

(c) Use the codes to

(i) draw a portion of the polyester molecule corresponding to the condensation of three different monomers. (1)

(ii) attach a styrene molecule to the polyester molecule at a place where cross-linking occurs. (1)

19. ***The Nitrogen Cycle***

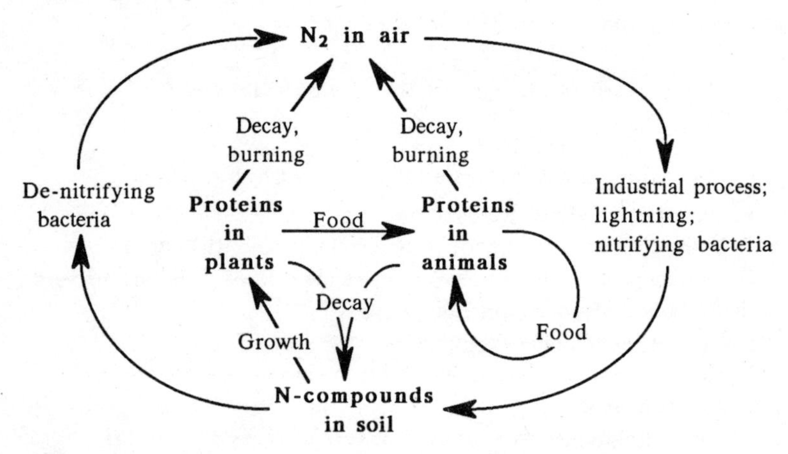

(a) (i) Name the industrial process referred to in the diagram. (1)

(ii) Write a chemical equation for the reaction caused by lightning. (1)

(iii) Describe briefly how some plants work with nitrifying bacteria. (2)

(b) Name the term that describes the processes in (a)(i), (ii) and (iii). (1)

(c) From the diagram state how (i) plants, (ii) animals acquire protein. (2)

20. In the Kjeldahl determination of flour protein, a known mass of flour is digested with concentrated sulphuric acid in the presence of selenium dioxide as catalyst.

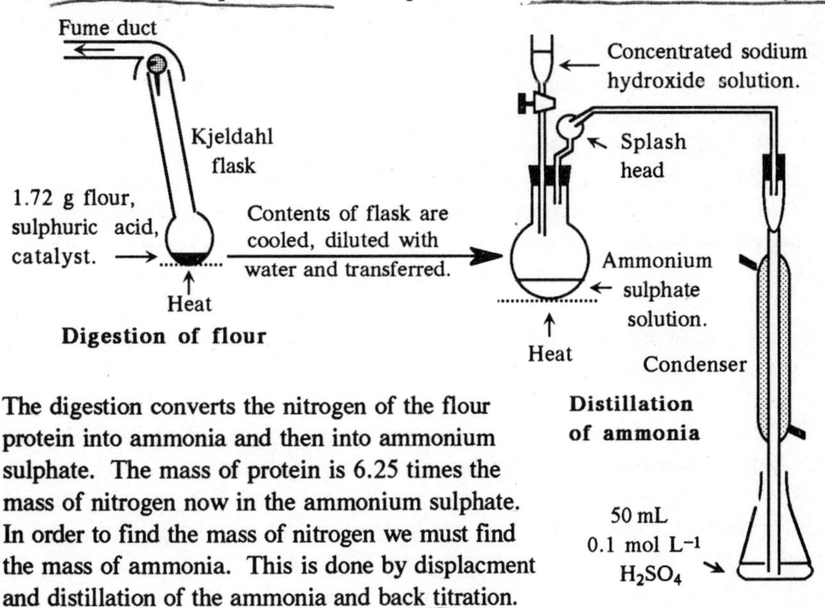

The digestion converts the nitrogen of the flour protein into ammonia and then into ammonium sulphate. The mass of protein is 6.25 times the mass of nitrogen now in the ammonium sulphate. In order to find the mass of nitrogen we must find the mass of ammonia. This is done by displacment and distillation of the ammonia and back titration.

(a) Write an equation for the reaction that occurs when sodium hydroxide solution is added to ammonium sulphate solution. (1)

(b) What is the purpose of the splash head? (1)

(c) Explain the term 'back titration'. (3)

(d) (i) What is the percentage nitrogen in flour protein (gluten)? (1)

(ii) The protein:nitrogen ratio for milk protein (casein) is 6.38:1. Is casein richer or poorer in nitrogen than gluten? (1)

Ammonia produced by the reaction in (a) distils over into the conical flask where it reacts with some of the sulphuric acid.

$$2NH_3(aq) + H_2SO_4(aq) \rightarrow (NH_4)_2SO_4(aq)$$

Titration of the residual acid in the flask shows that 38 mL of 0.1 mol L^{-1} acid remain after absorption of ammonia is complete.

(e) Name a reagent with which to titrate the residual acid in the flask. (1)

(f) State the volume of 0.1 mol L^{-1} sulphuric acid used up by the ammonia. (1)

(g) Calculate

(i) the mass of ammonia absorbed; (3)

(ii) the mass of nitrogen in the absorbed ammonia; (2)

(iii) the percentage protein in the flour. (2)

21. There are hundreds of natural amino acids. Glycine, NH_2CH_2COOH, is the simplest of the 20 or so <u>amino acids</u> that occur in <u>proteins</u>.

(a) Explain the underlined words. (2)
(b) Draw the full structural formulae of
(i) a dipeptide of glycine; (1)
(ii) a tripeptide of glycine; (1)
(iii) a portion of a polypeptide chain showing three glycine residues; (1)
(iv) the peptide group and use an arrow to indicate the position of the peptide bond. (1)
(c) Name the reaction that occurs when glycine molecules form a polypeptide. (1)

22. Write a short structural formula for a compound that is both basic and acidic and has the molecular formula $C_2H_5NO_2$. (1)

23. Write equations for the reactions between alanine, $NH_2CH(CH_3)COOH$, and (i) sodium hydroxide, (ii) hydrochloric acid, (iii) methanol. (3)

24. A fragment of a protein molecule was hydrolysed and chromatographed against eight known amino acids. Chromatogram 1 was obtained.

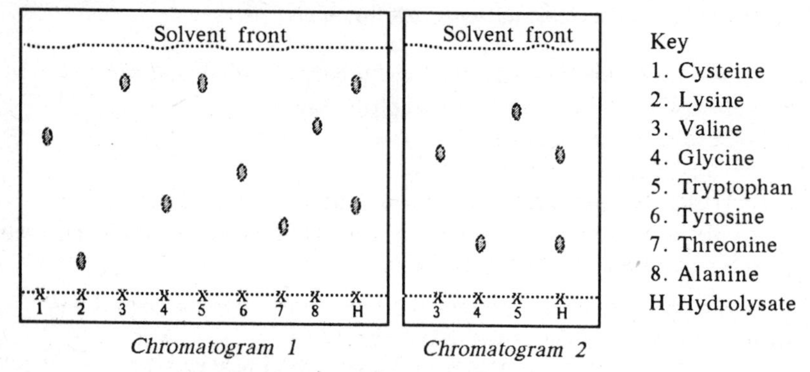

Chromatogram 1 *Chromatogram 2*

But Chromatogram 1 was inconclusive. The hydrolysate was then chromatographed against three of the eight amino acids, using a different solvent. Chromatogram 2 was obtained.

(a) State clearly the conclusions that may be drawn from Chromatogram 1. (2)
(b) Why was the chromatography repeated with
(i) acids 3 and 5; (1)
(ii) acid 4? (1)
(c) From Chromatogram 2, name the amino acids in the hydrolysate. (1)

25. Glycine (Gly), NH_2CH_2COOH, alanine (Ala), $NH_2CH(CH_3)COOH$ and cysteine (Cys), $NH_2CH(CH_2SH)COOH$ can give a variety of tripeptides. Show by drawing short formulae that the tripeptides Gly–Ala–Cys and Cys–Ala–Gly are not the same. (4)

26. "The biological role of chymotrypsin is to catalyze the hydrolysis of proteins in the small intestine.... Chymotrypsin does not cleave all peptide bonds.... Rather, it is selective for the peptide bonds on the carboxyl side of the aromatic side chains of tyrosine, tryptophan and phenylalanine." *BIOCHEMISTRY* by Lubert Stryer. Copyright (c) 1981 by Lubert Stryer. Reprinted with permission of W.H. Freeman and Company.

A phenylalanine residue in a protein chain.

(a) What is chymotrypsin called because of its biological role? (1)

(b) Explain why chymotrypsin is selective. (3)

(c) What becomes of the amino acids produced in the small intestine? (2)

(d) The body cannot make the benzene ring and so phenylalanine must be supplied in food. How is such an amino acid described? (1)

(e) Copy the structural formula of the residue shown above and
(i) draw a box around the aromatic side chain; (1)
(ii) identify the bond cleaved by chymotrypsin. (1)

(f) What structural feature do tyrosine, tryptophan and phenylalanine have in common? (1)

27. The hydrolysis of urea, a neutral substance, is catalysed by urease:

$$(NH_2)_2CO + H_2O \xrightarrow{\text{Urease}} 2NH_3 + CO_2$$

Exp.	T/°C	t/s
1.	12	5.5
2.	23	2
3.	32	1
4.	42	2
5.	55	9.5

The pH of the solution increases as hydrolysis progresses (due to the ammonia produced). Several hydrolyses were carried out at different temperatures. In each case the time taken for the pH to rise by the same 2 units was noted. Data are shown in the table.

Use these data to determine the optimum temperature for urease. (3)

28. The complexity of protein is almost infinite. Imagine a protein molecule composed of only 2 different acids, A and B: there are 2 possibilities (AB and BA) = 2 x 1. For 3 acids, A, B and C, there are 6 possibilities = 3 x 2 x 1. For 4 acids, A, B, C and D, there are 24 possibilities = 4 x 3 x 2 x 1. How many possibilities are there for 20 different amino acids? (2)

29. A chiral (pronounced kyral) molecule is one in which a central carbon atom is attached to four different groups. Because of this, such a molecule could exist in two forms, one the mirror image of the other, as shown opposite for alanine. But only one of these exists in proteins.

Naturally occurring alanine and its synthetic mirror image are chemically identical *except when they react with other chiral molecules.* Enzymes are chiral and react only with the natural alanine.

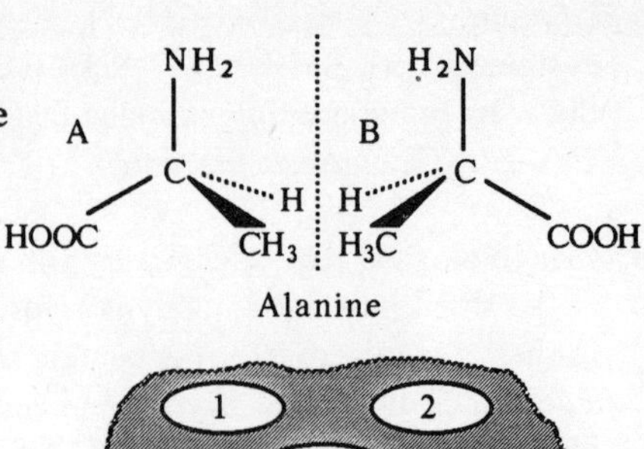

Receptor sites on enzyme.

(a) Explain why glycine is not chiral. (1)

(b) The receptors 1, 2 and 3 on the enzyme are specific for amino, carboxyl and methyl groups respectively.

(i) State whether A or B is the naturally occurring alanine. (1)

(ii) Explain, with the aid of a drawing, your choice in (i). (2)

30. Variable biochemical specificity is not restricted to chiral molecules. *Cis-trans* isomers may also vary in this respect. Reproduction in the silkworm moth depends on the male tracking a minute amount of the sex attractant (pheromone), bombykol, released by the female. Bombykol has the short formula:

$$CH_3(CH_2)_2CH{=}\overset{12}{C}H{-}CH{=}\overset{10}{C}H(CH_2)_8CH_2OH$$

Clearly, there are four possible isomers depending on the *cis-trans* configurations. The 10-*trans*, 12-*cis* isomer, for example, has the shape shown opposite. All four isomers were prepared and their biological activities measured.

OH

Outline shape of the 10-*trans*,12-*cis* isomer.

Compound	Biologically active concentration g cm^{-3}
Natural bombykol	10^{-16}
10-*trans*, 12-*cis* isomer	10^{-18}
10-*cis*, 12-*trans* isomer	10^{-9}
10-*cis*, 12-*cis* isomer	10^{-6}
10-*trans*, 12-*trans* isomer	10^{-5}

Reprinted with permission from 'The Shapes of Organic Molecules', by N. G. Clark, published by John Murray, London.

(a) Draw the outline shapes of the three other isomers. (3)

(b) Allowing for impurites in the natural material, state which of the isomers is bombykol? (1)

(c) Calculate the number of molecules per cm^3 for the most active isomer at its biologically active concentration. (3)

(d) By what factor is the most active greater than the least? (1)

31. The denaturation of a protein is shown schematically opposite. Globular proteins, such as egg albumin, milk casein and enzymes, all of which give colloidal solutions, are easily denatured.

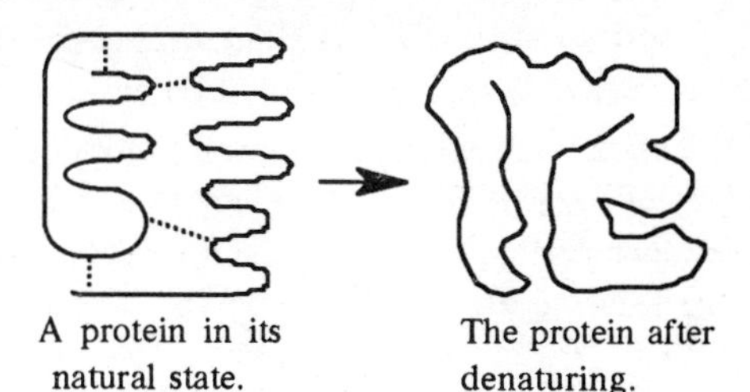

(a) Explain the underlined words. (3)

(b) During denaturation

(i) what happens to the physical structure of the protein; (1)

(ii) are peptide bonds broken? (1)

(c) Name two agents and/or conditions that may cause denaturation. (1)

32. Rennin is a proteolytic enzyme that coagulates milk; it produces a gel, which, on standing, separates into curds (a white solid) and whey (an aqueous supernatant liquid).

A student, who had determined rennin's optimum temperature as 38 °C, proposed to determine its optimum pH. She set up the experiment shown opposite for which data are shown in the table. Lactic acid was added to tubes 1 and 2 to lower the pH, sodium hydroxide solution to 4 and 5 to raise the pH.

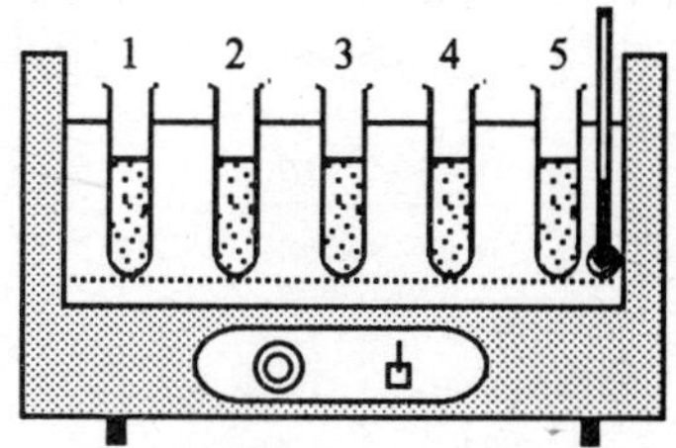

Each tube contains 10 mL milk, 0.5 mL rennin solution. Added acid (1, 2) and alkali (4, 5) varies the pH.

Tube	Acid /drops	Alkali /drops	Water /drops	pH of contents of tube	Time for curd to form /s
1	3	0	1	5.3	12
2	1	0	3	6.1	22
3	0	0	4	6.6	29
4	0	3	1	7.5	55
5	0	4	0	8.8	>300

(a) Explain the underlined words. (4)

(b) State the pH of milk. (1)

(c) Suggest a temperature at which the experiments should be carried out. (1)

(d) Why are drops of water added to tubes 1-4? (1)

(e) (i) Draw a graph of pH against time; (3)

(ii) Estimate the optimum pH for rennin; (1)

(iii) Say why the results are inconclusive. (1)

(f) Which industry is based on the changes in milk caused by rennin. (1)

UNIT 5
FROM BONDS TO BEHAVIOUR

1. Xenon is an <u>unreactive</u> gas but is not <u>inert</u>. It reacts with fluorine to form several fluorides. The formula of one of these fluorides was found from the following data. 0.249 g of the fluoride was decomposed into xenon and fluorine. The fluorine was absorbed, leaving 30 mL of xenon at r.t.s.p. R.a.m. of xenon = 131. V_m of xenon at r.t.s.p. = 24.8 L mol^{-1}.

   (a) (i) Which family of elements does xenon belong to? (1)
   (ii) Explain the underlined words. (2)
   (iii) Name an element in the same family that is inert. (1)
   (b) Calculate the masses of xenon and fluorine in the sample. (2)
   (c) Calculate the empirical formula of xenon fluoride. (2)
   (d) What additional information is needed to determine the molecular formula? (1)
   (e) Write an equation for the decomposition of xenon fluoride assuming the empirical formula to be the molecular formula. (1)

2. ***The Halogens***

Element	State at r.t.
Fluorine	Gas
Chlorine	Gas
Bromine	Liquid
Iodine	Solid

The physical states of the four principal halogens are shown opposite. Chemical reactivity decreases down the group.

   (a) (i) Refer to your data book and state the b.p.s of fluorine and chlorine. (1)
   (ii) Which would be more easily liquefied? (1)
   (b) Name and describe the bonding
   (i) within a fluorine molecule; (2)
   (ii) between fluorine molecules. (2)
   (c) Explain trends down the group in
   (i) physical state; (1)
   (ii) chemical reactivity. (2)

3. Lewis formulae are a convenient way of illustrating the covalent bond. The chlorine molecule, for example, which contains two chlorine atoms joined by a single covalent bond, is illustrated more readily by a Lewis formula than by a conventional orbit diagram.

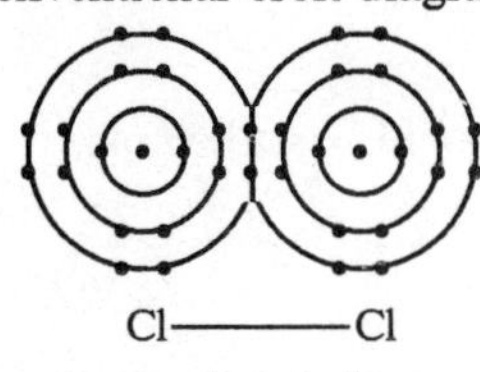

Conventional diagram.

:Cl : Cl:

Lewis formula.

The electron shells involved in bond formation are the only shells shown. (The electrons from different atoms may be distinguished by drawing them as dots and crosses.)

(Contd.)

Draw Lewis formulae for (a) H_2, (b) HF, (c) HCl, (d) HBr, (e) F_2, (f) Br_2, (g) I_2, (h) CH_4, (i) SiH_4, (j) CCl_4, (k) $SiBr_4$, (l) NH_3, (m) PH_3, (n) PI_3, (o) H_2O, (p) H_2S, (q) O_2, (r) N_2, (s) CO_2, (t) C_2H_6, (u) C_2H_4. (16)

4. Copper is a soft metal, but when mixed with a little tin to give bronze, it becomes much harder.

(a) What is a metal/metal mixture like bronze called? (1)

(b) Explain why bronze is harder than copper. You may find the data book helpful for this. (2)

5.

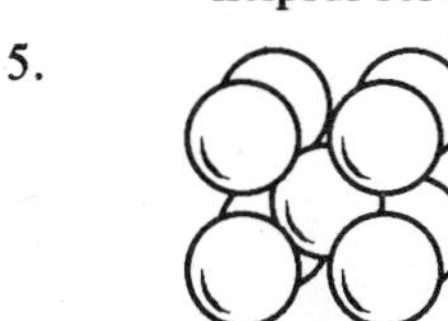

Potassium

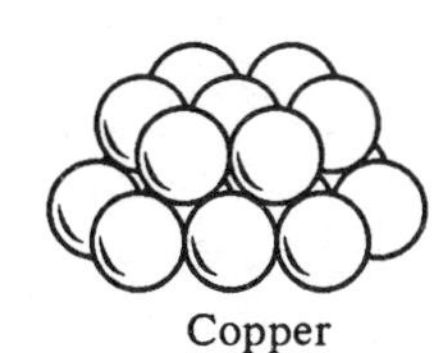

Copper

Potassium and copper are metals in the same row of the Periodic Table. The diagrams show how their respective atoms pack in the solid state.

(a) What general comments may be made about the chemical properties of elements in the same row of the Periodic Table? (2)

(b) Use your data book to find and state the densities of potassium and copper. (2)

(c) (i) By referring to the above diagrams, offer two reasons to explain the difference in density. (2)

(ii) State a third reason. You may find the data book helpful for this. (1)

6. Groups

1	2	3	4	5	6	7	8
A							B
C			D			E	
				F			
						G	
H							I

The short-form pattern of the Periodic Table is shown opposite. The places occupied by some elements are marked by letters. **The letters are not symbols.** Answer the following questions by selecting the appropriate letter or letters.

Which element

(a) is the lightest gas; (1)

(b) is the densest gas; (1)

(c) reacts with H to give the most ionic compound; (1)

(d) has a valency of 3; (1)

(e) may exist as a covalent network solid (giant molecule)? (1)

Which two elements

(f) consist of cations (positive ions) held together by mobile electrons; (1)

(g) are monatomic gases; (1)

(h) consist of discrete diatomic molecules; (1)

(i) react to give an interhalogen (halogen with halogen) compound? (1)

7. Six elements A to F selected from the first two rows of the Periodic Table

Li	Be	B	C	N	O	F	Ne
Na	Mg	Al	Si	P	S	Cl	Ar

are shown below.

A 

B 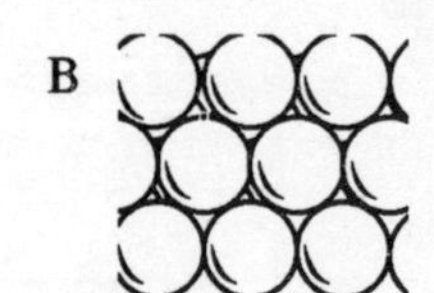

Covalent radius = 136 pm

C 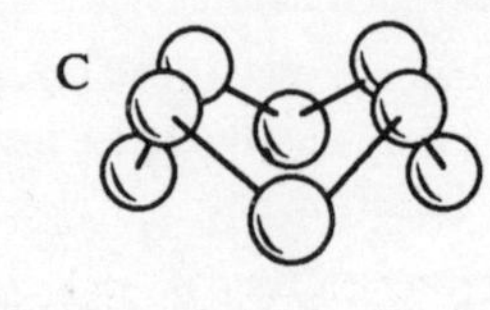

D 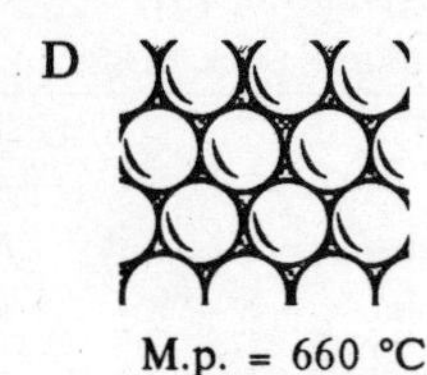

M.p. = 660 °C

E 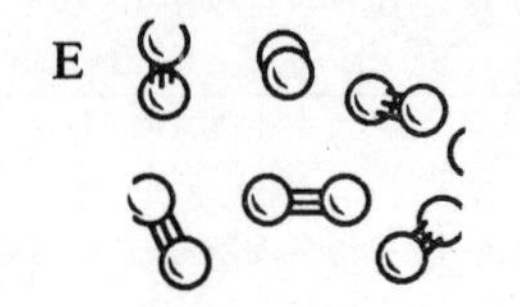

F 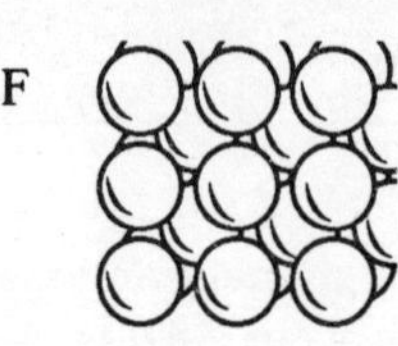

Density = 0.53 g cm^{-3}

(a) Identify each of the elements. (6)

(b) Explain why A is (i) greasy, (ii) a good electrical conductor. (2)

(c) Sketch the shape of a molecule of the element nearest to E in the same group and give its molecular formula. (2)

8.

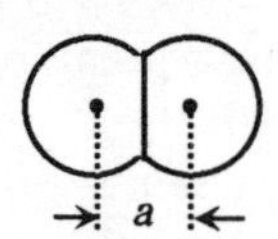

The distance *a* between chlorine nuclei in a Cl_2 molecule is 199 pm.

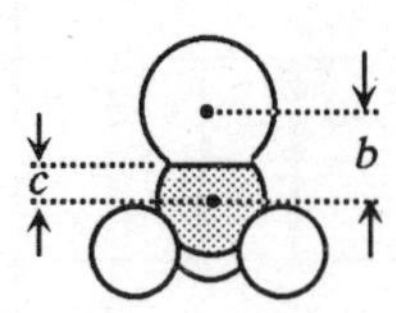

The distance *b* between carbon and chlorine nuclei in a CH_3Cl molecule is 177 pm.

(a) What are (i) $a/2$ and (ii) b called? (1)

(b) Calculate a value for *c*. (1)

9. The radii of the H^- and F^- ions are 208 and 136 pm respectively.

(a) Calculate the values of the ratios:

(i) $\frac{\text{Ionic radius of } H^-}{\text{Covalent radius of H}}$ (ii) $\frac{\text{Ionic radius of } F^-}{\text{Covalent radius of F}}$ (2)

(b) Given that hydrogen and fluorine atoms each gain one electron to become ions, why are the ratios so different? (2)

10.

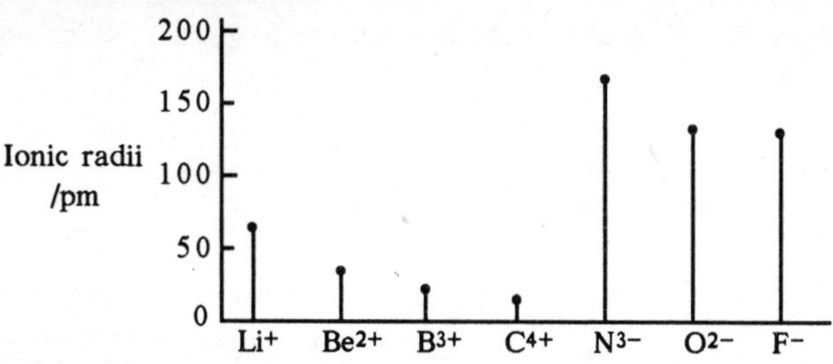

(a) Explain why
(i) the ionic radii decrease from Li^+ to C^{4+}; (1)
(ii) there is a marked increase in ionic radii from C^{4+} to N^{3-}; (1)
(iii) the ionic radii decrease from N^{3-} to F^-. (1)

(b) The information in this question is displayed as a spike graph. Why is this preferable to a line graph or a bar graph? (2)

11. The table below gives the ionic radii of some particles.

Particle	O^{2-}	F^-	Ne	Na^+	Mg^{2+}	Al^{3+}
Radius/pm	140	136	-	95	65	50

(a) In what way do the nuclei of each of these particles differ? (1)
(b) What do the particles have the same? (1)
(c) What would your answer to (b) lead you to expect about the radii of each of these particles? (1)
(d) Offer a reason to account for the decrease in size from left to right. (1)
(e) (i) Why is there no value for neon? (1)
(ii) From the ionic radii, what should be the radius of a neon atom? (1)

12.

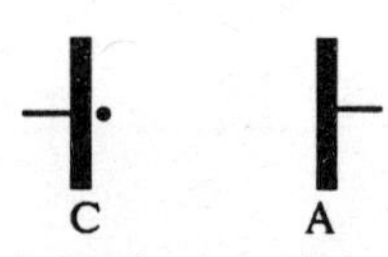

A stationary electron (on C) has no kinetic energy.

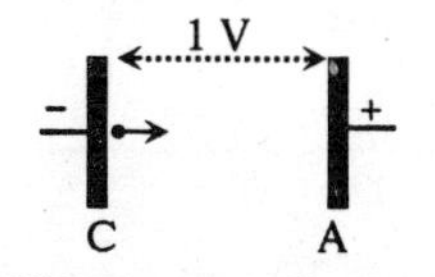

When a potential of 1 V is applied to the plates, the electron accelerates away from C.

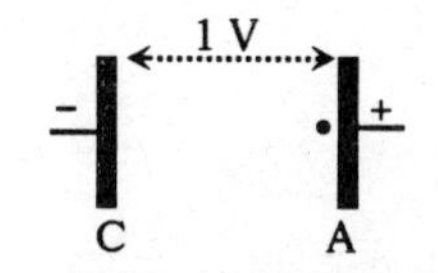

The kinetic energy of the electron just as it hits A is 1.602×10^{-19} J.

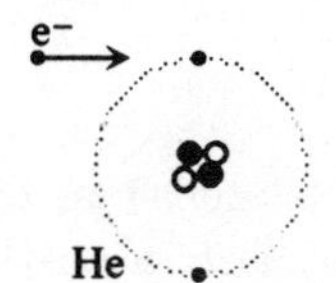

(a) What would be the kinetic energy of the electron if accelerated (i) through 5 V, (ii) through 8.9 V? (2)
(b) Name the kinetic energy needed by the accelerated electron if it were to be able to knock an electron off a helium atom? (1)

(Contd.)

(c) A potential of 24.6 V is needed to give an electron enough kinetic energy to knock an electron off a helium atom.
(i) Write an equation for the reaction. (1)
(ii) Calculate the energy change per one atom of helium. (1)
(iii) Calculate the energy change per mole of helium. (1)

13. Ionisation energies for elements P, Q, R and S are shown in the table below.

Element	Ionisation Energies /kJ mol^{-1}		
	1st	2nd	3rd
P	720	1460	3100
Q	1320	3400	5320
R	420	3060	4440
S	1260	2310	3840

(a) (i) Which element belongs to Group 1 of the Periodic Table? (1)
(ii) Explain your answer to (i). (2)
(b) The energy change for

$$E(g) \rightarrow E^{3+}(g) + 3e^-$$

is 7410 kJ mol^{-1}. Identify E from the table (1)

14. The 1st and 2nd ionisation energies for the elements H to Ne are shown below.

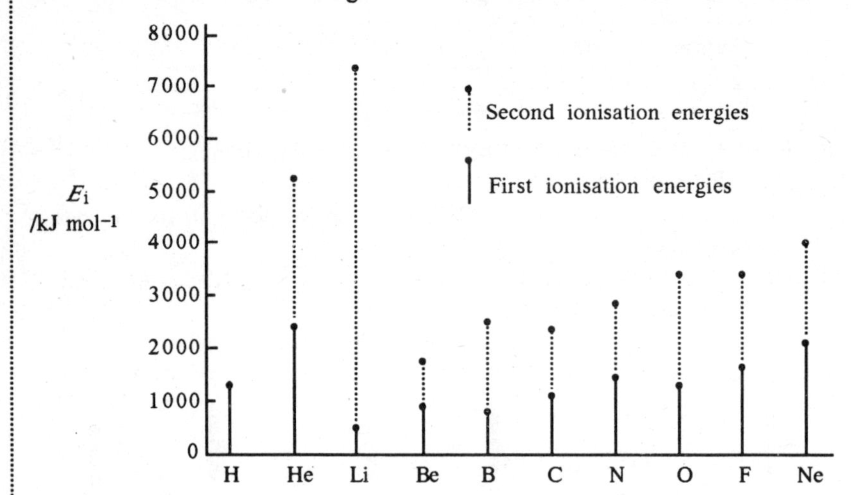

(a) Explain the change in first ionisation energy
(i) from H to He; (2)
(ii) from He to Li; (2)
(iii) from Li to Be. (1)
(b) Explain the change in second ionisation energy from He to Li. (1)
(c) Why is there no second ionisation energy for hydrogen? (1)
(d) Why is the second ionisation energy for each element greater than the first? (1)
(e) Which element has a third ionisation energy much greater than its second? (1)
(f) Explain the *general* trend in ionisation energies from Be to Ne. (1)

15. 

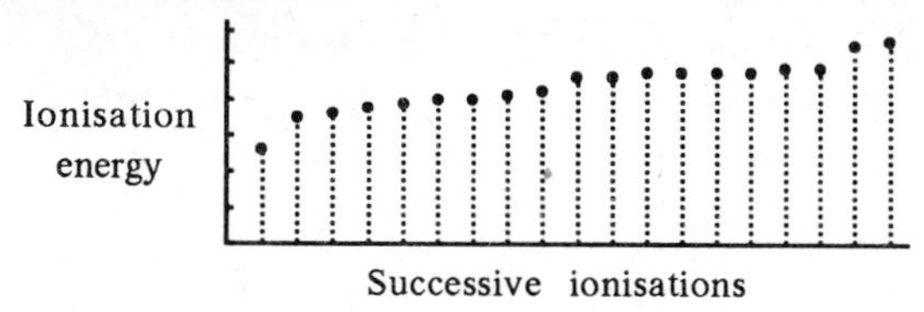

The graph shows the successive ionisations of an element down to the nucleus. Give the electron configuration of the element. (1)

16. ***The Alkali Metals***

Li 2.1
Na 2.8.1
K 2.8.8.1
Rb 2.8.18.8.8.1
Cs 2.8.18.18.8.1
(Fr 2.8.18.32.18.8.1)

The electron configurations of the alkali metals are shown opposite. These metals are malleable and ductile. They have similar but not identical reactions: they all displace hydrogen from water, for example, but the rates at which they do so increase rapidly down the group.

(a) Which group of the Periodic Table do these metals belong to? (1)
(b) Explain the underlined words. (2)
(c) In terms of electron configurations, explain why their reactions are (i) similar but (ii) not identical. (3)
(d) Select any two alkali metals and show by writing chemical equations that their reactions with water are similar. (2)
(e) Why does reactivity increase down the group? (2)

17. Unequal sharing of bonding electrons leads progressively from covalency through polar covalency to electrovalency, though no bond ever achieves 100% electrovalency.

(a) Arrange the substances listed below in order of increasing electrovalency.
Sodium chloride (NaCl), ammonia (NH_3), lithium iodide (LiI), caesium fluoride (CsF), hydrogen fluoride (HF), hydrogen (H_2). (2)
(b) Explain the meaning of the formula $H^{\delta+} — Cl^{\delta-}$. (2)
(c) What is the approximate percentage ionic (electrovalent) character of the most ionic substance in (a) above. (1)

18.

Element	Formula of hydride	Action of hydride on moist pH paper
Sodium	(i)	(ii)
Carbon	(iii)	(iv)
Nitrogen	(v)	(vi)
Chlorine	(vii)	(viii)

Write answers for (i) to (viii). (4)

19\.

Li Be B C N O F
Na Mg Al Si P S Cl
K Ca Br
Rb Sr I
Cs Ba

Give ionic/covalent hydrides.
Give covalent hydrides.
Give ionic hydrides.

The hydrides of the elements may be classified in three ways according to bond type.

(a) What determines bond type? (3)
(b) (i) In the Na row, which hydride is most covalent? (1)
(ii) Explain your choice. (2)

20\. The ionic hydrides are white crystalline solids prepared by direct synthesis. Lithium hydride is the only one that can be melted without decomposition; caesium hydride in fact ignites spontaneously when exposed to air.

(a) (i) Explain the underlined words. (1)
(ii) Sketch an apparatus that could be used to make lithium hydride. (3)
(b) Does thermal stability of the hydride increase or decrease down the group? (1)
(c) Write an equation for the spontaneous ignition of caesium hydride in air. (1)

21\. When lithium hydride is dissolved in molten lithium chloride at 650 °C and electrolysed, hydrogen gas is set free at electrode Y as shown opposite.

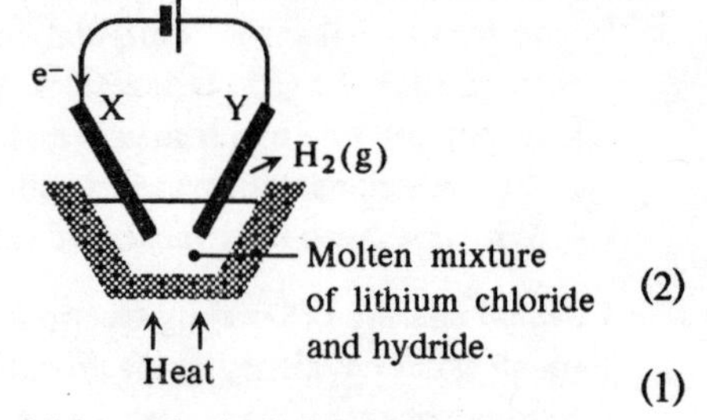

(a) Write ion-electron equations for the half-reactions occurring at X and Y. (2)
(b) What is unusual about the evolution of hydrogen at Y? (1)
(c) Why is the electrolysis of a solution of lithium hydride in water not an alternative? Write an equation in support of your answer. (2)

22\.

Li Be B C N O F
Na
K
Rb
Cs
+e−
−e−
Cs^+ F^-

Atoms of lithium row and Group 1 elements are shown opposite. These are drawn from covalent atomic radii. The comparative sizes of Cs^+ and F^- ions are also shown.

(a) Explain the changes in atomic size
(i) across the row from left to right; (1)
(ii) down the group. (1)
(b) Why is no atom shown for neon? (1)
(c) Explain why
(i) the caesium ion is smaller than the caesium atom; (1)
(ii) the fluoride ion is larger than the fluorine atom. (1)

23. Radius ratio is the value of the quotient

$$\frac{\text{Radius of smaller ion}}{\text{Radius of larger ion}}$$

Radius ratios may be used to predict how ions will usually pack together. Radius ratio ranges are shown in the table. The way ions pack in caesium fluoride is shown opposite.

Radius ratio ranges	No. of ions around each oppositely charged ion
0.155–0.225	3 (triangular structure)
0.225–0.414	4 (tetrahedral structure)
0.414–0.732	6 ("NaCl structure")
>0.732	8 ("CsCl structure")

Caesium fluoride lattice

(a) Which kind of lattice is shown opposite? (1)

(b) (i) Calculate the radius ratio of CsF. (Radii of Cs^+ and F^- are 169 and 136 pm.) (1)

(ii) From your answer to (i) state the structure you expect for CsF. (1)

(iii) Examine the CsF lattice and state which structure it is. Explain your choice. (2)

(iv) Is the radius ratio prediction for CsF correct? (1)

24. (a) Draw diagrams that show the three-dimensional shapes of the PH_3 and CCl_4 molecules. (2)

(b) The P—H bond is non-polar yet the PH_3 molecule is polar; the C—Cl bond is polar yet the CCl_4 molecule is non-polar. Explain these apparent contradictions. (2)

25. Water is a polar liquid: it consists of polar molecules and is attracted strongly to a charged surface. Non-polar liquids are attracted only slightly or not at all. Explain why

(i) a water molecule is polar; (1)

(ii) a water molecule is attracted to a charged surface; (1)

(iii) a polar molecule is not repelled from a charged surface; (1)

(iv) non-polar molecules experience slight attraction? (1)

26. There are three dichloroethenes, all of which are liquids at r.t.:

```
1.  H   H      2.  H   Cl      3.  H   Cl
    |   |          |   |           |   |
    C = C          C = C           C = C
    |   |          |   |           |   |
    Cl  Cl         Cl  H           H   Cl
```

(a) In each case, state whether the liquid would be polar or non-polar. (3)

(b) Describe an experiment and sketch the apparatus to test your answers to (a). (3)

(c) (i) Which liquid would you expect to have the lowest b.p.? (1)

(ii) Explain your choice in (i). (1)

27. CH_2ClCH_2Cl (1,2-dichloroethane) is non-polar but is strongly attracted to a charged surface. Explain. (3)

28. The diagram shows a plot of b.p. against molecular mass for alcohols, straight alkanes and branched alkanes.

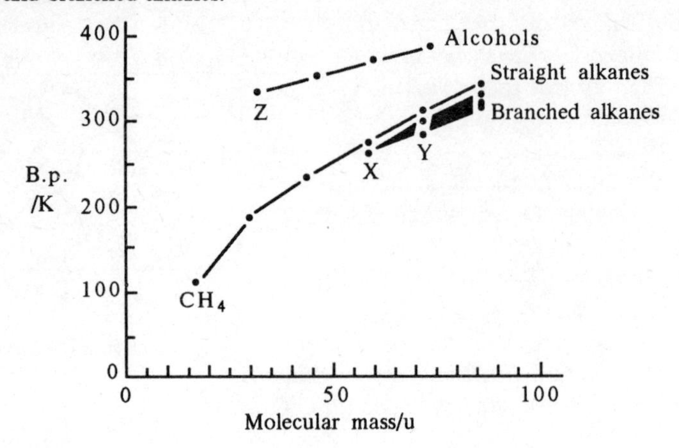

(a) Explain the change in b.p. through the straight alkane series. (1)
(b) Draw a full structural formula for X. (1)
(c) (i) For a given molecular mass, do the branched alkanes boil at a higher or lower temperature than the straight alkanes? (1)
(ii) Offer a reason for the difference in b.p. observed in (i). (2)
(iii) In view of your answer to (ii), draw a short structural formula for Y. (1)
(iv) Explain why the branched line does not start at C_3. (1)
(d) State the name and formula of Z. (1)
(e) Explain the difference in b.p.s between alcohols and alkanes. (3)

29. When hydrogen chloride dissolves in water, it gives a conducting solution, but a non-conducting solution in benzene. Explain this difference in behaviour. (2)

30. Explain why water, which consists of small H_2O molecules, is more viscous than pentane, which consists of large $CH_3CH_2CH_2CH_2CH_3$ molecules. (2)

31. The viscosity of a mixture of chloroform, $CHCl_3$, and ether, $C_2H_5OC_2H_5$, varies with the composition of the mixture as shown in the opposite graph.

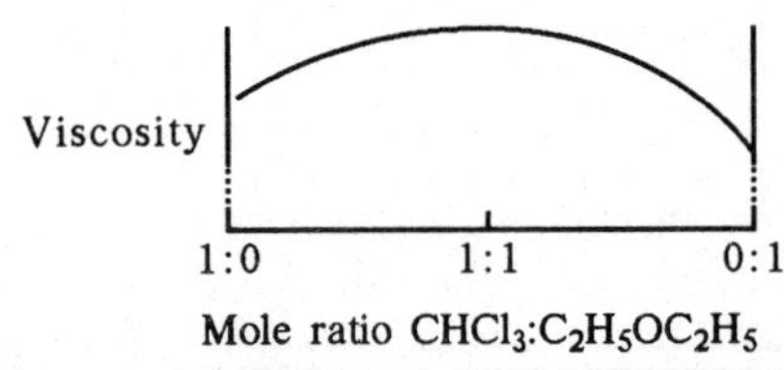

(Contd.)

(a) What is the meaning of 'viscosity'? (1)
(b) Which pure liquid has the lower viscosity? (1)
(c) At what composition does the mixture have maximum viscosity? (1)
(d) Explain why the mixture is more viscous than either pure liquid alone. (2)

32. A student asserted that he could predict the state of a substance at r.t. from a knowledge of molecular mass, using the alkanes as reference:

Gas → $C_4H_{10}(g)$ 58 u → Liquid → $C_{17}H_{36}(l)$ 240 u → Solid →

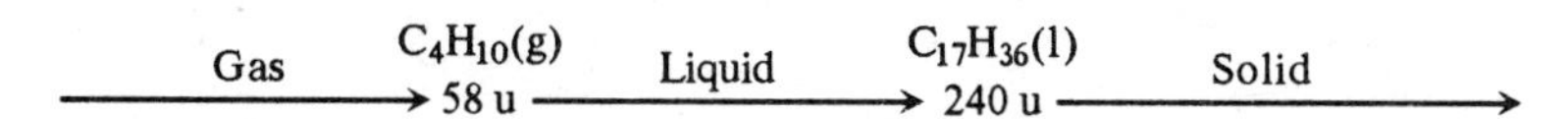

Anything with a molecular mass less than 58 u would be a gas, and so on. When he applied his theory to the substances (i) iodine, (ii) sulphur, (iii) carbon dioxide, (iv) silicon dioxide, (v) water and (vi) hydrogen sulphide, he achieved only partial success.

(a) Take each in turn and say whether his prediction is right or wrong. (6)
(b) If wrong, why is it wrong? (2)

33. The lattices of two minerals are shown below. The data in pm are covalent radii.

117 pm
74 pm
74 pm
37 pm

Mineral A Mineral B

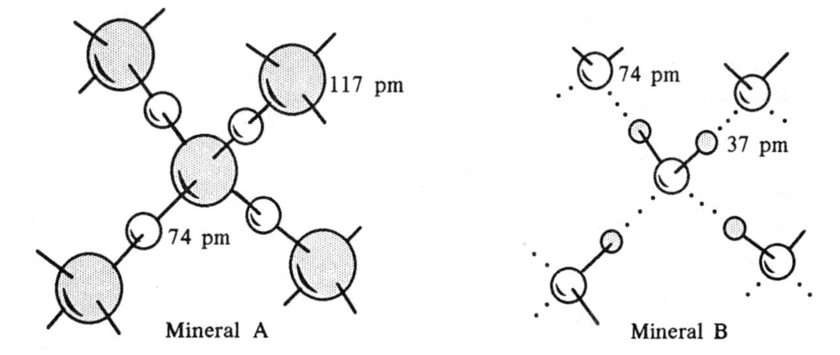

(a) Use information in the data book to name A and B. (2)
(b) State the type of lattice in A and in B. (2)
(c) Explain why the m.p.s of the two lattices are very different. (2)
(d) (i) Name two other substances that adopt the same type of lattice as A. (1)
(ii) What physical property other than m.p. do they have in common? (1)

34.

Air at –30°C
Ice at –30 °C
Ice at 0 °C
Water at 0 °C
Water at 4 °C

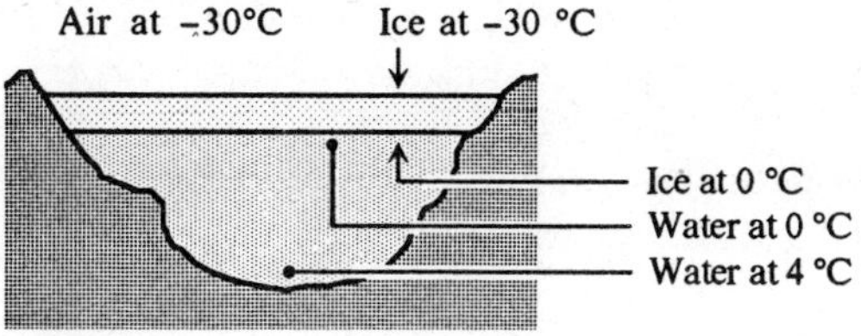

The profile of a highland loch in winter is shown opposite. Explain how hydrogen bonds are responsible for this profile. (3)

35. 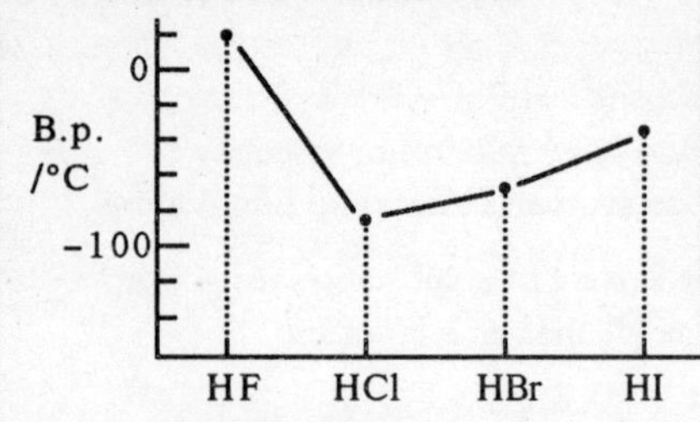

The graph illustrates the anomalous b.p. of hydrogen fluoride.

(a) What does 'anomalous' mean? (1)

(b) Why is the b.p. of hydrogen fluoride anomalous, and what would be the b.p. if it were not anomalous? (1)

(c) (i) What is responsible for the anomalous behaviour of hydrogen fluoride? (2)

(ii) Show by drawing formulae how your explanation in (i) accounts for the presence of species with molecular masses of 40 and 60 u. (1)

36. The molecular masses of methane and ammonia are almost equal, yet methane boils at a much lower temperature (109 K) than ammonia (240 K).

(a) State the molecular masses of (i) methane and (ii) ammonia. (1)

(b) Why do we expect molecules of similar mass to have similar b.p.s? (1)

(c) Why are the b.p.s of these substances very different? (2)

37. Propanoic acid in the vapour state is partially associated to give dimers; ethanoic acid crystals contain infinite chains of associated molecules.

```
        O···H–O
      //        \
C2H5–C           C–C2H5
      \        //
        O–H···O
```

Dimer of propanoic acid

```
     CH3          CH3          CH3
      |            |            |
···O=C–O–H ···O=C–O–H ···O=C–O–H···
```

Ethanoic acid chain

(a) Explain the underlined words. (2)

(b) Name the bond that acts between these molecules. (1)

(c) (i) Describe how this bond holds these molecules together. (2)

(ii) State three criteria (conditions) for the bond to form. (3)

38. Lithium iodide, caesium fluoride, hydrogen chloride, hydrogen oxide, gold, neon, carbon (diamond); silicon carbide; carbon (graphite); silicon dioxide; fluorine.

Select a substance (or substances)

(a) in which the bonding is 100% covalent;

(b) composed of polar covalent molecules;

(c) that has an elevated boiling point because of hydrogen bonding;

(d) in which the bonding is van der Waals, covalent and metallic;

(e) that is held together by delocalised mobile electrons only;

(f) that is the most ionic;

(g) in which the bonding is van der Waals only;

(h) that consists of layers held together by van der Waals forces;

(i) that consists of diatomic molecules held together by van der Waals forces. (7)

39. Write a formula for the salt produced when the following amphoteric oxides react with sodium hydroxide: (a) Al_2O_3; (b) ZnO, (c) PbO_2, (d) PbO, (e) VO_2, (f) As_2O_3, (g) SnO, (h) Sb_2O_3. (4)

40.

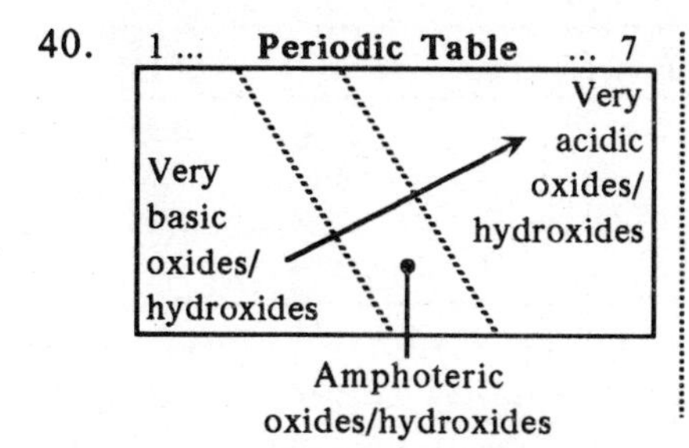

(a) Explain the trend in acid-base character of oxides/hydroxides shown in the table. (3)
(b) What is an amphoteric oxide? (1)
(c) Select any one amphoteric oxide and write balanced chemical equations to illustrate its amphoteric behaviour. (2)
(d) Name an organic amphoteric substance. (1)

41. Lithium iodide is soluble in both water and non-aqueous solvents while caesium fluoride is soluble in water only. What does this information convey about the bonding in these salts? (2)

42. ***Dissolving a salt***

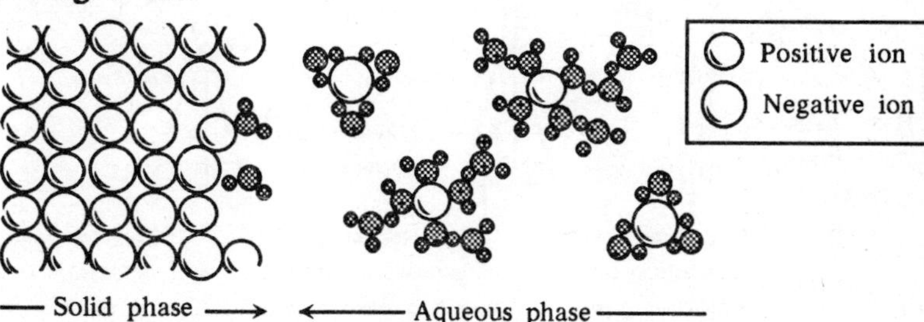

Water molecules become attached to ions as they quit their lattice positions. The ions and their water molecules are then free to wander about in solution.

(a) The diagram shows that water molecules have a preference in the way they attach to ions. State this preference and offer a reason for it. (2)
(b) Ions move at different speeds. It might seem reasonable for a small ion to move faster than a large ion, but the opposite is often true. What is the reason for this? (1)
(c) Describe the energy changes that occur as the salt dissolves. (3)

43. State the type of reaction in each of the following.

(a) $K^+Br^-(s) \xrightarrow{\text{Water}} K^+(aq) + Br^-(aq)$ (1)
(b) $SiCl_4(l) + 2H_2O(l) \rightarrow SiO_2(s) + 4HCl(g)$. (1)
(c) $CuSO_4(s) + 5H_2O(l) \rightarrow CuSO_4.5H_2O(s)$. (1)
(d) $HCl(g) \xrightarrow{\text{Water}} H^+(aq) + Cl^-(aq)$. (1)
(e) $K(g) \rightarrow K^+(g) + e^-$ (1)

44. Water <u>hydrolyses</u> phosphorus trichloride, <u>dissociates</u> sodium chloride, <u>ionises</u> hydrogen chloride and <u>hydrates</u> copper sulphate. Explain the underlined words and write equations to support your explanations. (8)

45. Standard reduction electrode potentials of the halogens are shown in the table.

Half-reaction	$E^{\ominus}/V$
1. $I_2(s) + 2e^- \rightleftharpoons 2I^-(aq)$	0.54
2. $Br_2(l) + 2e^- \rightleftharpoons 2Br^-(aq)$	1.07
3. $Cl_2(g) + 2e^- \rightleftharpoons 2Cl^-(aq)$	1.36
4. $F_2(g) + 2e^- \rightleftharpoons 2F^-(aq)$	2.85

(a) Which species is (i) the strongest oxidiser, (ii) the weakest oxidiser, (iii) the strongest reducer, (iv) the weakest reducer? (4)

(b) Use the potentials to arrange the halogens in decreasing order of reactivity. (1)

46. In order to test predictions based on the reduction potentials of the halogens (Question 45), a student set up the experiment shown below.

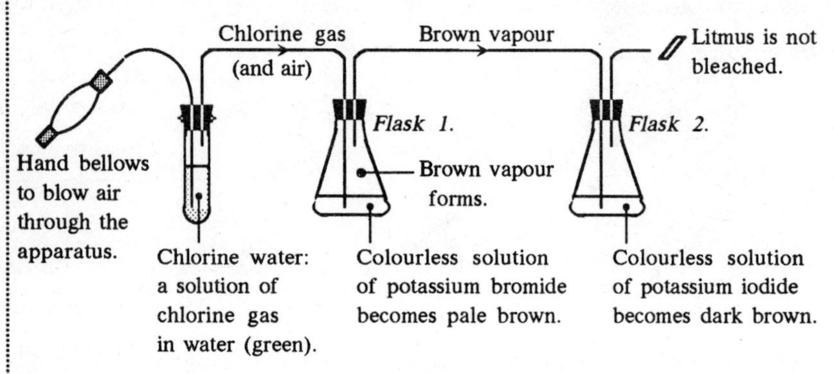

Flask 1

(a) Name the brown vapour. (1)

(b) (i) Write equations for the half-reactions occurring in the flask. (2)

(ii) State which half-reaction is a reduction and which an oxidation. (1)

(c) (i) Write an ionic equation for the reaction occurring in the flask. (1)

(ii) What type of reaction is it? (1)

Flask 2

(d) (i) What substance is responsible for the colour that develops. (1)

(ii) Describe a test for this substance. (2)

(e) Write an ionic equation for the reaction occurring in the flask. (1)

(f) What is the significance of the litmus test? (2)

UNIT 6
THERMOCHEMISTRY

1. To suction pump
Thermometer
Stirrer
Air at 20.3 °C
Copper spiral
Calorimeter
Burner
Air enters here.

The heat of combustion of methanol can be found using the apparatus shown opposite.

The calorimeter contains 800 g of water at 20.3 °C. The initial mass of the burner and the methanol it contains is 8.345 g. The final mass is 7.933 g. The final temperature of the water is 22.9 °C.

(a) What is the purpose of the spiral? (2)

(b) Why should the temperature of the water not be allowed to rise more than about 3 °C? (1)

(c) Write an equation for the combustion of methanol. (1)

(d) Calculate

(i) the heat absorbed by the water. (c = 4.18 kJ K^{-1} kg^{-1}.) (1)

(ii) the heat of combustion, ΔH, of methanol. (2)

(e) (i) What is the data book value for this ΔH. (1)

(ii) Offer a reason for the difference between this value and that in (d)(ii). (1)

2. The apparatus of question 1 can be modified to determine the heats of combustion of foods. The burner is replaced with a nickel crucible containing a known mass of the food, electrical ignition is used and oxygen is substituted for air.

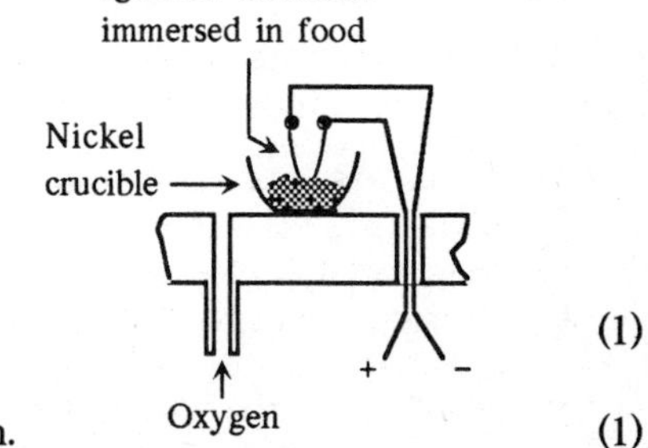

(a) Why is oxygen used instead of air? (1)

(b) Describe what you would expect to see when the ignition current is switched on. (1)

(c) (i) In view of your answer to (b), state a correction that should be made to the determined heat of combustion. (1)

(ii) How would you obtain a value for this correction? (1)

(d) What term is popularly used to describe the heat of combustion of a food? (1)

3. An average human weighs 65 kg and generates 10 500 kJ of heat per day due to metabolic activity. What would be the temperature rise in one day if none of this heat were lost. (Take c for the body as 4.18 kJ K^{-1} kg^{-1}.) (2)

4. The graph shows how the enthalpy change, ΔH, for the reaction:
$N_2(g) + 3H_2(g) \rightarrow 2NH_3(g)$
varies with temperature.
State the value of ΔH for the reaction:
$NH_3(g) \rightarrow \frac{1}{2}N_2(g) + 1\frac{1}{2}H_2(g)$
at 600 K. (2)

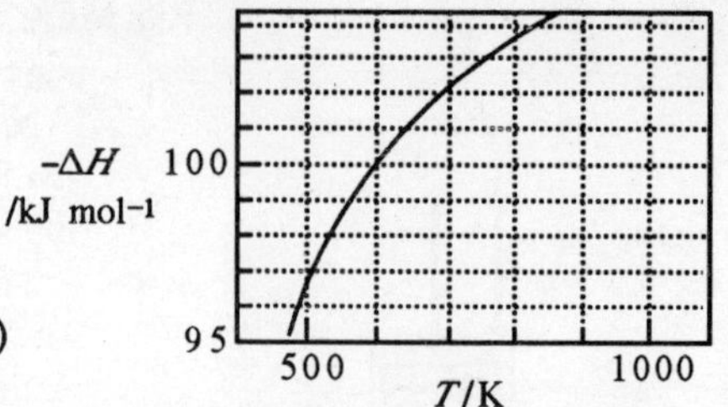

5. ***Heat of neutralisation***

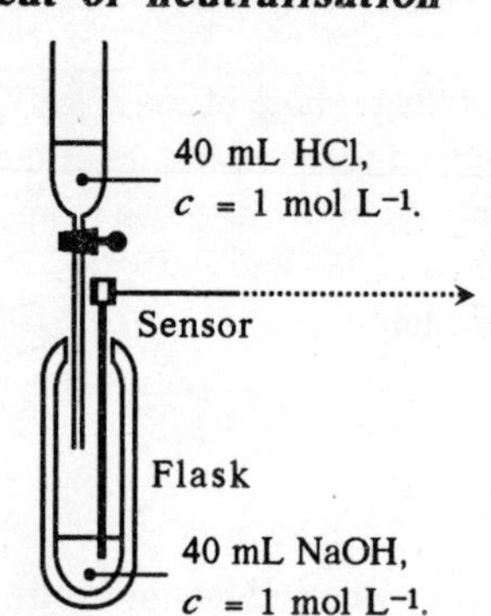

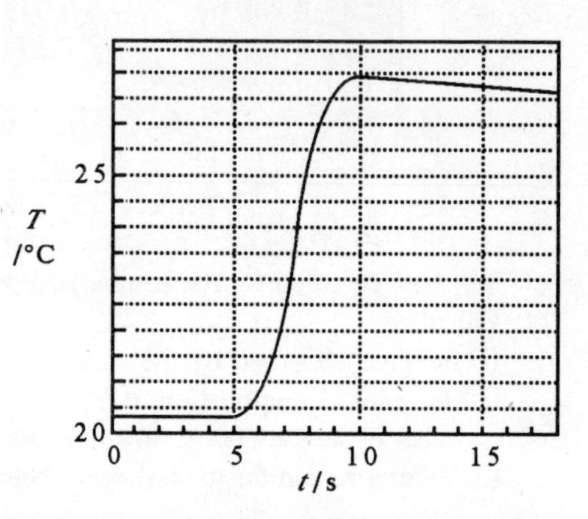

The heat of neutralisation of hydrochloric acid and sodium hydroxide was found using the above apparatus. A sensor connected to a computer was used to give a print-out of temperature during the neutralisation. The acid and alkali each had the same initial temperature.

(a) What was the initial temperature? (1)
(b) When was the addition of acid (i) started, (ii) finished? (1)
(c) Use the graph to determine ΔT for the neutralisation. (1)
(d) Calculate the heat of neutralisation. (c is given in the data book.) (3)

6. In this question, a solution whose concentration is 1 mol L^{-1} will be called a standard solution: symbol = S.
When 25 mL of S strong acid is mixed with 25 mL of S strong base, the rise in temperature is 7 °C. State the temperature rise in each of the following.

(a) 50 mL of S HCl with 50 mL of S NaOH.
(b) 100 mL of 2 S HCl with 100 mL of 2 S NaOH.
(c) 80 mL of S HCl with 40 mL of S KOH.
(d) 30 mL of 0.5 S HCl with 30 mL of 0.5 S NaOH.
(e) 60 mL of S HCl with 60 mL of 0.5 S NaOH.
(f) 40 mL of S H_2SO_4 with 40 mL of 2 S KOH. (6)

7. 0.64 g of ammonium nitrate was dissolved in 100 g of water. The temperature decreased by 0.5 K. Use $cm\Delta T$ to calculate the enthalpy of solution of the salt. (c = 4.2 kJ K^{-1} kg^{-1}.) (3)

8. Some heats of hydration Ion(g) → Ion(aq) are shown in the table. Which two factors determine the value of ΔH_{hyd}? Illustrate your answer with examples from the table. (3)

Ion	$-\Delta H$ hyd. kJ mol^{-1}	Ion	$-\Delta H$ hyd. kJ mol^{-1}	Ion	$-\Delta H$ hyd. kJ mol^{-1}
Li^+	519				
Na^+	406	Mg^{2+}	1920	Al^{3+}	4690
K^+	322	Ca^{2+}	1650		
Rb^+	301	Sr^{2+}	1480		
Cs^+	276	Ba^{2+}	1360		

9. The enthalpy of formation of carbon dioxide, CO_2, is ΔH = −394 kJ mol^{-1}, and that of dinitrogen monoxide, N_2O, is ΔH = +74 kJ mol^{-1}.

   (a) (i) Which of the compounds CO_2 or N_2O is an endothermic compound? (1)
   (ii) How would the other one be described? (1)

   (b) Write chemical equations for
   (i) the enthalpies of formation of CO_2 and N_2O; (2)
   (ii) the combustion of carbon in (excess) dinitrogen monoxide. (1)

   (c) Calculate the enthalpy change for the reaction in (b)(ii). (2)

   (d) What assumption is made about the enthalpies of formation of carbon, oxygen and nitrogen, and of elements in general? (1)

   (e) Draw
   (i) a spike graph of the heats of formation of CO_2 and N_2O; (2)
   (ii) an enthalpy diagram for the reaction in (b)(ii). (3)

10. Write chemical equations for the enthalpies of formation of (a) H_2O, (b) CH_4, (c) C_3H_6, (d) NH_4NO_3, (e) C_2H_5OH, (f) $C_6H_{12}O_6$, (g) $C_{16}H_{18}O_4N_2S$. (7)

11.

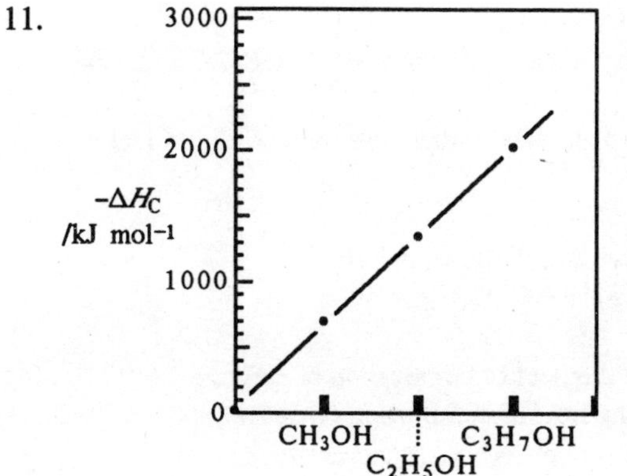

The graph gives the heats of combustion of three alcohols: CH_3OH, CH_3CH_2OH $CH_3CH_2CH_2OH$.

   (a) Why is the line straight? (2)

   (b) Use the graph to estimate the heat of combustion, ΔH_C, of butanol. (1)

   (c) Why should the line pass through the origin? (2)

12. ***The dissolution of potassium iodide.***

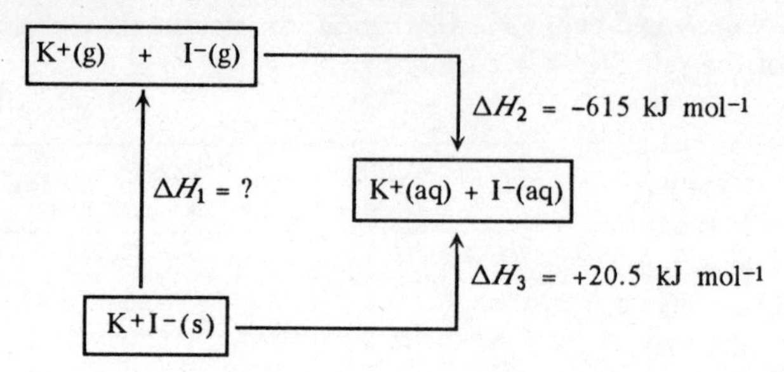

(a) Name the three enthalpy changes ΔH_1, ΔH_2 and ΔH_3. (3)

(b) (i) Write an algebraic expression for ΔH_1, ΔH_2 and ΔH_3. (1)

(ii) Use this expression to calculate a value for ΔH_1. (1)

(iii) Does temperature rise or fall when potassium iodide is dissolved in water? (1)

(c) Given that ΔH for $K^+(g) \rightarrow K^+(aq)$ is -322 kJ mol^{-1}, calculate a value for $I^-(g) \rightarrow I^-(aq)$. (1)

13. Lattice, hydration and solution enthalpies for three fluorides are given in the table but x, y and z are omitted.

Calculate the values of x, y and z. (3)

Fluoride	$\Delta H_{latt.}$ /kJ mol^{-1}	$\Delta H_{hyd.}$ /kJ mol^{-1}	$\Delta H_{soln.}$ /kJ mol^{-1}
LiF	x	−1017	+5
NaF	+902	y	0
KF	+801	−819	z

14. RDX, $C_3H_6N_6O_6$, is a powerful explosive. It is a white crystalline solid. Its heat of formation is $\Delta H_f = +60$ kJ mol^{-1}. It decomposes according to:

$$C_3H_6N_6O_6(s) \longrightarrow 3CO(g) + 3H_2O(g) + 3N_2(g);\ \Delta H = -1100 \text{ kJ mol}^{-1}.$$

(a) From the information given above state three reasons why RDX might be a good explosive. (3)

(b) State a fourth factor, not mentioned above, that ensures that it is in fact a very powerful explosive. (1)

15. Carbon burns in excess oxygen according to the equation:

$$C(s) + O_2(g) \longrightarrow CO_2(g).$$

(a) Use bond energies to calculate the heat of combustion of carbon. (4)

(b) State the data book value for the heat of combustion of carbon. (1)

(c) Offer one reason for the difference between your answers to (a) and (b). (1)

16. 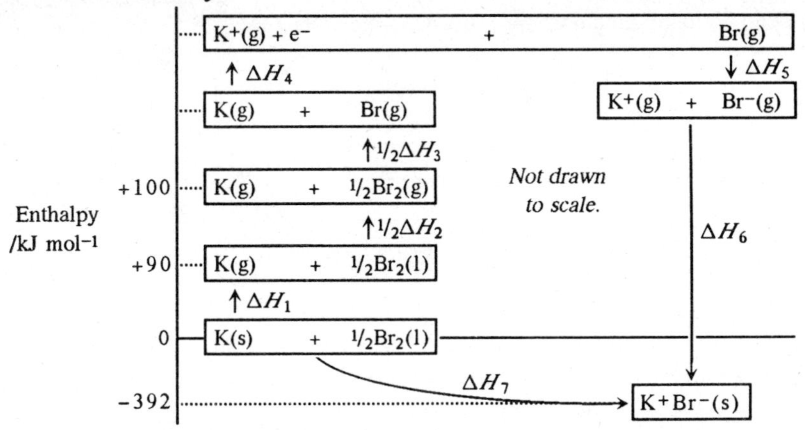

The bond enthalpy diagram shown above is called a Born-Haber cycle. This particular cycle illustrates the individual steps in the formation of potassium bromide from its elements at r.t.s.p..

(a) Write an expression that relates the seven enthalpy changes in the diagram. (1)
(b) Identify ΔH_7 and state its value. (1)
(c) Construct a table with the headings:

ΔH	Name of enthalpy change	Value/kJ mol^{-1}

Complete the table for ΔH_1, ΔH_2, ΔH_3 and ΔH_4. (4)
(d) ΔH_5 is called the electron affinity of bromine. It has the value –342 kJ mol^{-1}. Write a chemical equation for this reaction. (1)
(e) Calculate ΔH_6, the lattice energy of potassium bromide. (2)

17. Use bond energies in the table to calculate the heat of formation of nitrogen trifluoride (a gas at r.t.s.p.). State whether nitrogen trifluoride is an exothermic or endothermic compound. (4)

Bond	Bond energy/kJ mol^{-1}
N≡N	949
F – F	155
N – F	255

18. (a) (i) Calculate ΔH for the reaction

$$H_2(g) + \tfrac{1}{2}O_2(g) \longrightarrow H_2O(g)$$

from bond energies in the data book. (3)
(ii) Why is the value for this ΔH *not* the heat of formation of water? (1)
(b) When 1 mol of steam condenses and cools to r.t.s.p.,

$$H_2O(g) \longrightarrow H_2O(l),$$

47 kJ of heat are evolved. State the heat of formation of water. (1)

19. A bond enthalpy diagram* for propene is shown below.

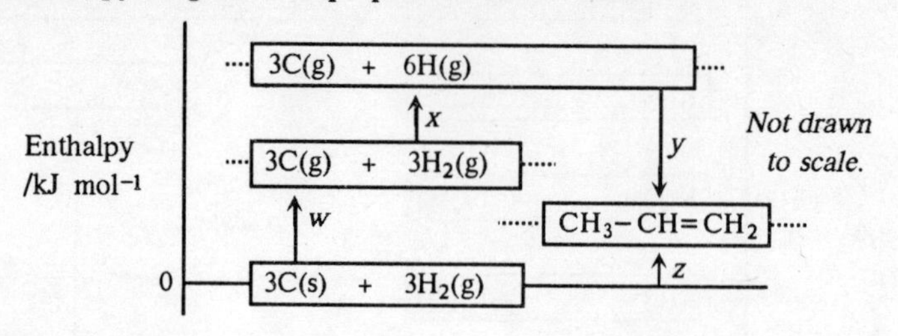

(a) What is z called? (1)

(b) Use the data book to find values for w, x and y, and hence determine z. (4)

**Do not confuse the '3C(g) + 6H(g)' of a bond enthalpy diagram with the activated complex of an enthalpy diagram of Unit 1.*

20. A bond enthalpy diagram for the formation of boron trifluoride is shown below.

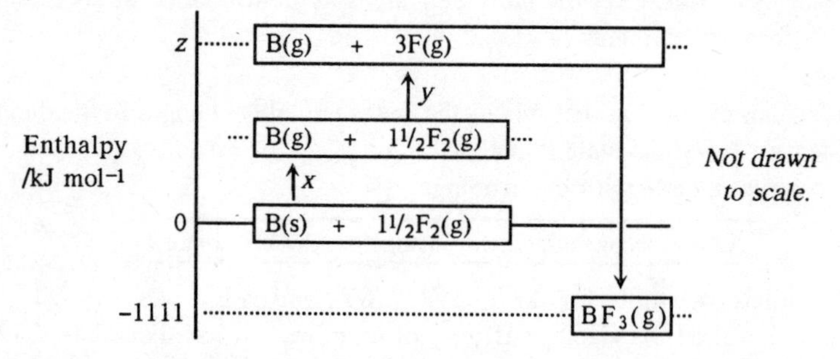

The mean B—F bond energy is 645 kJ mol^{-1}.

(a) Refer to your data book and state the value for y. (1)

(b) Calculate x, the heat of sublimation of boron. (2)

(c) State the value of z. (1)

21. $Be(s) + Cl_2(g) \longrightarrow BeCl_2(s)$; $\Delta H_f = -512$ kJ mol^{-1}

The mean Be—Cl bond energy is 389 kJ mol^{-1}.

$\Delta H_{sublimation}$ of Be is 321 kJ mol^{-1}.

Use this information and the Cl—Cl bond energy in the data book to find the sublimation energy of $BeCl_2$. (4)

22. Cyclopropane is a saturated hydrocarbon that behaves as if it were unsaturated:

```
  H   H
  |   |                          H  H  H
H-C———C-H                        |  |  |
   \ /       + HBr  ——>        H-C--C--C-Br; ΔH = -3 kJ mol-1.
    C                            |  |  |
   / \                           H  H  H
  H   H
```

(Contd.)

(a) In what way does cyclopropane behave as if it were unsaturated? (2)

(b) (i) What is the C—C—C bond angle in cyclopropane? (1)

(ii) What is the approximate tetrahedral bond angle preferred by carbon? (1)

Tetrahedral angle

(iii) Offer a reason for the unsaturated property of cyclopropane. (1)

Bond energies offer an alternative reason: the more energy evolved when a bond forms, the more stable the bond.

(c) (i) Would you expect the C—C bond energy in cyclopropane to be greater or smaller than the C—C value in the data book? (1)

(ii) Explain your answer to (i). (1)

(d) State the data book C—C mean bond energy. (1)

(e) Use $\Delta H = -3$ kJ mol^{-1} and bond energies in the data book to calculate the C—C bond energy in cyclopropane. (3)

23. ***Hess's Law***

Two routes by which solid sodium hydroxide is converted into sodium ethanoate solution are shown in the diagram.

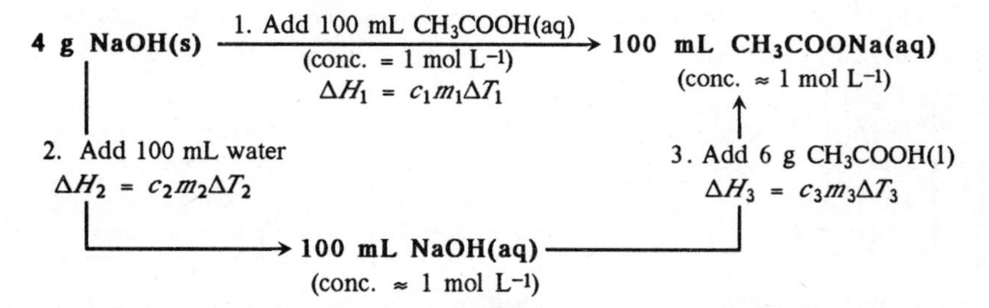

One route is direct, the other, indirect. The direct route consists of step 1. The indirect route consists of steps 2 and 3.

(a) Express Hess's law in terms of ΔH_1, ΔH_2 and ΔH_3. (1)

(b) c_1, c_2 and c_3 are approximately equal, as also are m_1, m_2 and m_3. Write an expression for Hess's law in terms of ΔT_1, ΔT_2 and ΔT_3. (1)

(c) Experimental results are given in the table. Show that these confirm Hess's law. (2)

Step	Temperature/°C Initial	Final
1	23.6	46.1
2	21.5	31.6
3	31.0	43.4

24.

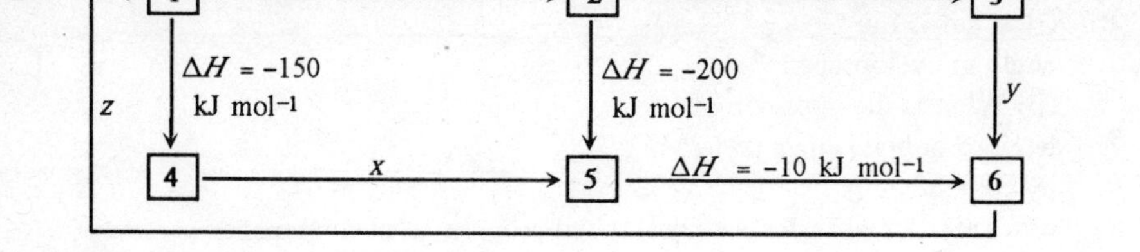

Solve for x, y and z. (3)

25. The diagram shows two enthalpy changes, ΔH_1 and ΔH_3, that can be measured by experiment. The other enthalpy change, ΔH_2, cannot be so determined but can be calculated from ΔH_1 and ΔH_3.

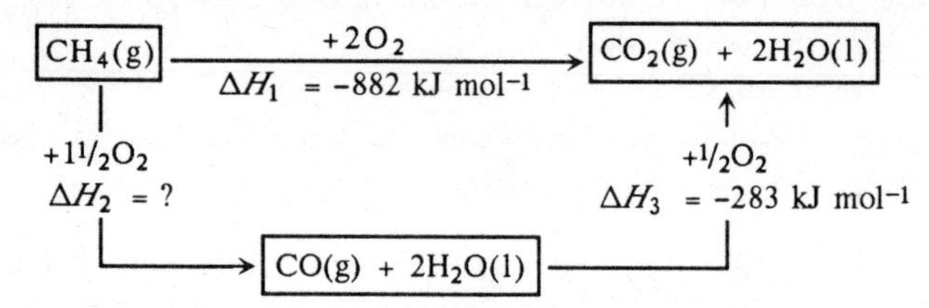

(a) State the type of reaction illustrated in the diagram. (1)

(b) Why cannot reaction 2 be carried out experimentally? (1)

(c) (i) Write an expression linking ΔH_1, ΔH_2 and ΔH_3. (1)

(ii) Name and state the law on which this expression is based. (2)

(iii) Calculate ΔH_2. (1)

26.

Bond breaking enthalpy changes (which include all endothermic steps).	+	*Bond making enthalpy changes (which include all exothermic steps).*	=	*Enthalpy change for the reaction as written in the chemical equation.*

The heat of formation of tetrachloromethane is the heat change for the reaction

$$C(s) + 2Cl_2(g) \longrightarrow CCl_4(l)$$

at r.t.s.p. and is $\Delta H_f = -139$ kJ mol^{-1}. Use this information together with the heat of sublimation of carbon and bond dissociation energy of chlorine given in the data book to calculate the enthalpy change for the (physical) reaction

$$CCl_4(g) \longrightarrow CCl_4(l).$$

(4)

27.

Equation	ΔH/kJ mol^{-1}
1. $Na(s) \rightarrow Na(g)$	+109
2. $Na(g) \rightarrow Na^+(g) + e^-$	+502
3. $Na^+(g) \rightarrow Na^+(aq)$	−406

(a) Identify reactions 1, 2 and 3. (3)

(b) Calculate ΔH for the reaction

$$Na(s) \longrightarrow Na^+(aq) + e^-$$

from the information in the table. (2)

28. A bond enthalpy diagram for the formation of ammonia is shown below.

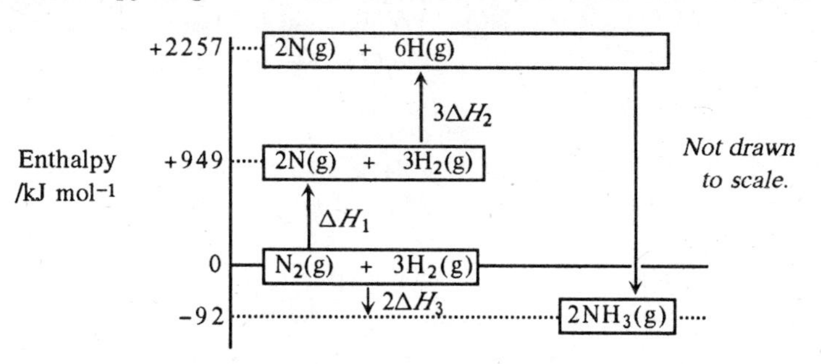

(a) Why are $N_2(g)$ and $3H_2(g)$ placed on the 0 of the enthalpy scale? (1)

(b) Name and evaluate ΔH_1, ΔH_2 and ΔH_3. (3)

(c) (i) Write an expression in ΔH_1, ΔH_2, N—H bond formation energy (let this = x) and ΔH_3. (1)

(ii) Calculate the N—H bond formation energy. (1)

29. The heat of sublimation of selenium is ΔH_s = +202 kJ mol^{-1}. The heat of formation of hydrogen selenide, H_2Se, is ΔH_f = +86 kJ mol^{-1}.

$$H_2(g) + Se(s) \longrightarrow H_2Se(g)$$

(a) Use this information and the H—H bond energy in the data book to build a bond enthalpy diagram for this reaction. (3)

(b) Use the diagram to find the H—Se bond dissociation energy. (1)

30. Use the heats of combustion of carbon, hydrogen and methanol in the table to calculate the heat of formation of methanol. (5)

Substance	ΔH_c /kJ mol^{-1}
Carbon	-394
Hydrogen	-286
Methanol	-715

31. Use the heats of formation of carbon dioxide, water and methanol in the table to calculate the heat of combustion of methanol. (5)

Substance	ΔH_f/kJ mol^{-1}
Carbon dioxide	-394
Water	-286
Methanol	-239

32. Calculate the mean O—H bond energy in the H_2O molecule from the data:

$$2H(g) \longrightarrow H_2(g); \quad \Delta H = -435 \text{ kJ mol}^{-1}$$
$$2O(g) \longrightarrow O_2(g); \quad \Delta H = -495 \text{ kJ mol}^{-1}$$
$$2H_2(g) + O_2(g) \longrightarrow 2H_2O(l); \quad \Delta H = -572 \text{ kJ mol}^{-1}$$
$$H_2O(g) \longrightarrow H_2O(l); \quad \Delta H = -41 \text{ kJ mol}^{-1}$$

(5)

33. Given: $2HCl(g) \longrightarrow 2H(g) + 2Cl(g); \Delta H = +862 \text{ kJ mol}^{-1}$
$H_2(g) \longrightarrow 2H(g); \Delta H = +436 \text{ kJ mol}^{-1}$
$Cl_2(g) \longrightarrow 2Cl(g); \Delta H = +243 \text{ kJ mol}^{-1}$

(a) State the H—Cl bond energy per mole. (1)
(b) (i) Write an equation for the formation of hydrogen chloride. (1)
(ii) Calculate the enthalpy of formation of hydrogen chloride. (2)

34. Given: $ICl(g) \longrightarrow I(g) + Cl(g); \Delta H = +203 \text{ kJ mol}^{-1}$
$I_2(s) \longrightarrow I_2(g); \Delta H = +37 \text{ kJ mol}^{-1}$
$I_2(g) \longrightarrow 2I(g); \Delta H = +161 \text{ kJ mol}^{-1}$
$Cl_2(g) \longrightarrow 2Cl(g); \Delta H = +243 \text{ kJ mol}^{-1}$

(a) State the I—Cl bond formation energy. (1)
(b) (i) Write an equation for the formation of iodine chloride (a gas at r.t.s.p.). (1)
(ii) Calculate the enthalpy of formation of iodine chloride. (3)

35. The electron affinity of chlorine is ΔH for the reaction
$$Cl(g) + e^- \longrightarrow Cl^-(g).$$
The dissociation of chlorine is
$$1.\ Cl_2(g) \longrightarrow 2Cl(g); \quad \Delta H = +243 \text{ kJ mol}^{-1}$$
The hydration of the chloride ion is
$$2.\ Cl^-(g) \longrightarrow Cl^-(aq); \Delta H = -364 \text{ kJ mol}^{-1}$$
The formation of the aqueous chloride ion is
$$3.\ \tfrac{1}{2}Cl_2(g) + e^- \longrightarrow Cl^-(aq); \Delta H = -624 \text{ kJ mol}^{-1}.$$
Use equations 1, 2 and 3 to determine the electron affinity of chlorine. (4)

36. State whether the general reactions listed below are exothermic, endothermic, or either. (E = element, M = metal, Nm = non-metal.)

(a) Combustion, e.g., $E + O_2 \rightarrow EO_2$.
(b) Dissociation (bond breaking), e.g., $Nm_2 \rightarrow 2Nm$.
(c) Dissolution, e.g., $MNm(s) + \text{water} \rightarrow MNm(aq)$.
(d) Lattice disruption, e.g., $MNm(s) \rightarrow M^+(g) + Nm^-(g)$.
(e) Electron affinity, e.g., $Nm(g) + e^- \rightarrow Nm^-(g)$.
(f) Boiling, evaporation, vaporisation : liquid → vapour.
(g) Fusion (melting): solid to liquid.
(h) Formation: elements → compound.
(i) Ionisation: $E(g) \rightarrow E^+(g) + e^-$.
(j) Neutralisation: $H^+(aq) + OH^-(aq) \rightarrow H_2O(l)$.
(k) Hydration, e.g., $M^+(g) + \text{water} \rightarrow M^+(aq)$.
(l) Sublimation: solid → gas.
(m) Atomisation: $M(s) \rightarrow M(g)$. (13)

UNIT 7
CHEMICAL EQUILIBRIUM

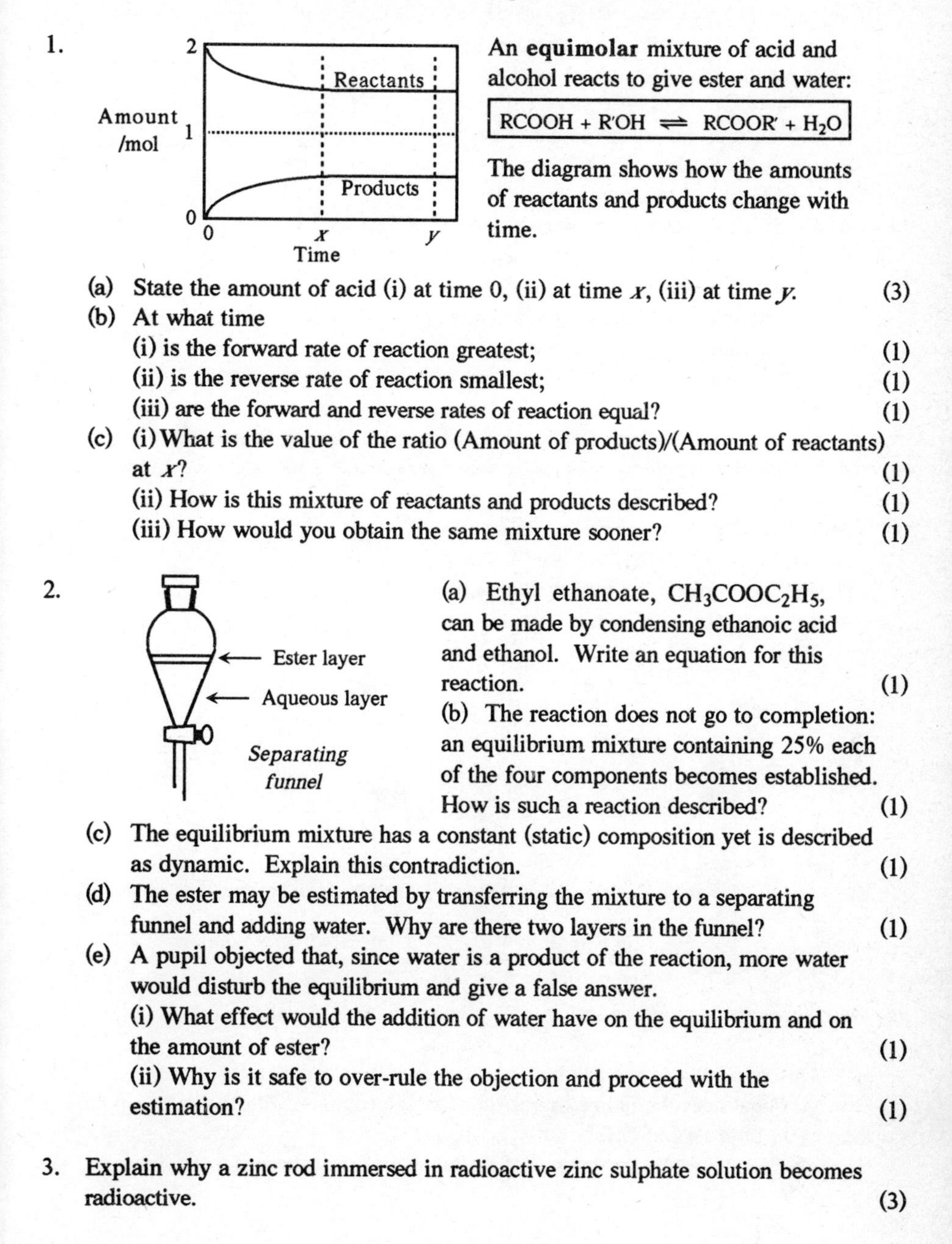

1. An **equimolar** mixture of acid and alcohol reacts to give ester and water:

$$RCOOH + R'OH \rightleftharpoons RCOOR' + H_2O$$

The diagram shows how the amounts of reactants and products change with time.

(a) State the amount of acid (i) at time 0, (ii) at time x, (iii) at time y. (3)

(b) At what time

(i) is the forward rate of reaction greatest; (1)

(ii) is the reverse rate of reaction smallest; (1)

(iii) are the forward and reverse rates of reaction equal? (1)

(c) (i) What is the value of the ratio (Amount of products)/(Amount of reactants) at x? (1)

(ii) How is this mixture of reactants and products described? (1)

(iii) How would you obtain the same mixture sooner? (1)

2. (a) Ethyl ethanoate, $CH_3COOC_2H_5$, can be made by condensing ethanoic acid and ethanol. Write an equation for this reaction. (1)

(b) The reaction does not go to completion: an equilibrium mixture containing 25% each of the four components becomes established. How is such a reaction described? (1)

(c) The equilibrium mixture has a constant (static) composition yet is described as dynamic. Explain this contradiction. (1)

(d) The ester may be estimated by transferring the mixture to a separating funnel and adding water. Why are there two layers in the funnel? (1)

(e) A pupil objected that, since water is a product of the reaction, more water would disturb the equilibrium and give a false answer.

(i) What effect would the addition of water have on the equilibrium and on the amount of ester? (1)

(ii) Why is it safe to over-rule the objection and proceed with the estimation? (1)

3. Explain why a zinc rod immersed in radioactive zinc sulphate solution becomes radioactive. (3)

4. Rate curves for the reforming of methane $CH_4(g) + H_2O(g) \rightleftharpoons CO(g) + 3H_2(g)$ are shown in the graph. r_f is the forward rate of reaction and r_b the backward rate.

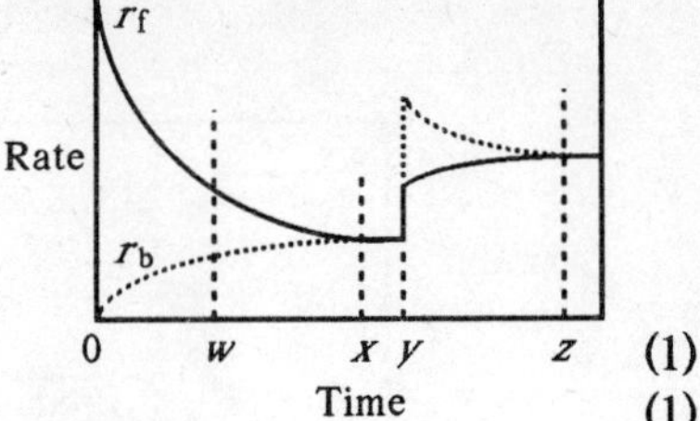

(a) Which substances are present in the reaction vessel at time 0? (1)
(b) Which rate is greater at time *w*? (1)
(c) What is the relation between r_f and r_b at time *x*? (1)
(d) Temperature is increased at time *y*.
(i) Which rate is increased more? (1)
(ii) Is the forward reaction exothermic or endothermic? (1)
(iii) Explain your answer to (ii). (2)
(iv) Compare the equilibrium composition at *z* with that at *x*. (1)

5. Explain why the addition of radioactive magnesium chloride to saturated non-radioactive magnesium chloride solution produces a radioactive solution. (3)

6. A salt crystal was immersed in saturated salt solution as shown in the diagram. Over a few hours the crystal (viewed through a lens) changed its shape as shown below.

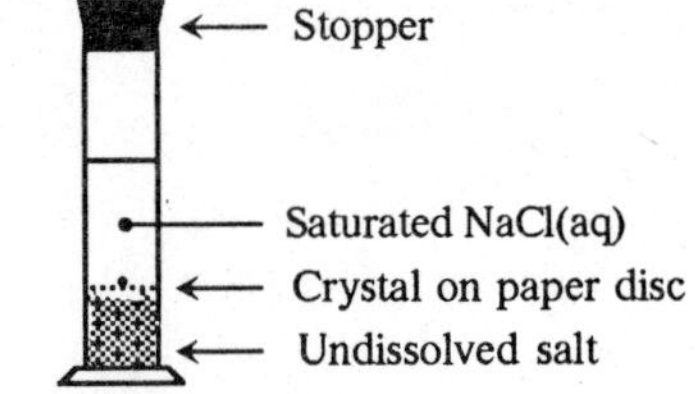

Initial shape → Shape after a few hours

Explain why this change of shape occurs. (3)

7.

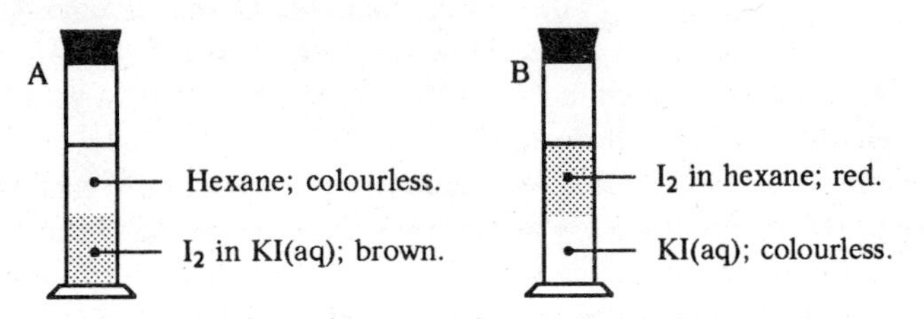

Two tubes were set up as shown using the same mass of iodine in each.

(a) Describe what is **seen** to take place in each tube over several days. (1)
(b) (i) What does the iodine in one tube have in common with the iodine in the other after several days? (1)
(ii) How does the appearance of the tubes confirm your answer in (i)? (1)

8. ***Ammonia synthesis: the Haber process***

Percentage NH_3 at equilibrium							
	Pressure/bar						
T/°C	1	10	50	100	200	300	600
350		7	25				
450		2	9	16			
550	0.08			7	12	36	53
650	0.03			3	6		
750	0.02			2	3		
850	0.01			1	2		
950	0.005				1		

Data with permission from Rational Approach to Chemical Principles by J. A. Cranston, published Blackie and Son Ltd.

(a) State how the equilibrium % NH_3 varies with (i) increasing temperature and (ii) increasing pressure. (1)

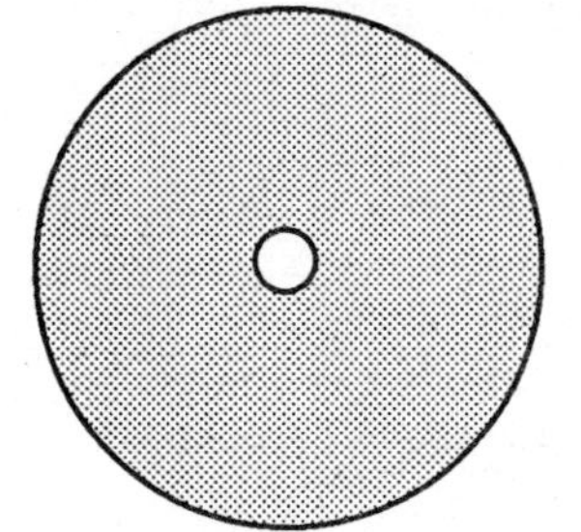

Cross-section, actual size, of a steel pipe used in the synthesis of ammonia.

(b) Estimate the equilibrium %NH_3 at 450 °C and 200 bars (the usual operating conditions in this country). (1)

(c) Ammonia is removed when the %NH_3 is about 15%. Account for the difference between this percentage and that in (b). (2)

(d) State one advantage and one disadvantage of using a higher temperature. (1)

(e) (i) Why does the pipe shown opposite have this unusual cross-section? (1)
(ii) State one disadvantage of using plant like this. (1)

9. When nitrogen monoxide, NO, is heated to 1000 °C, it decomposes completely into a mixture of nitrogen, N_2, and oxygen, O_2. The decomposition is exothermic. When this mixture is heated to 3000 °C, an equilibrium that contains about 5% nitrogen monoxide is formed.

(a) Why does nitrogen monoxide not decompose at 600°C? (1)

(b) (i) Write an equation for the N_2(g), O_2(g), NO(g) equilibrium and state (1)
(ii) whether the forward reaction is exothermic or endothermic, (1)
(iii) the effect of increasing the temperature, (1)
(iv) the effect of increasing the pressure. (1)

(c) There are 16000000 thunderstorms worldwide each year; and the temperature in a lightning flash is 30000 °C (hotter than the sun!). Explain why these storms are good for plant growth. (2)

10. The graph opposite shows the rate at which product P forms in a mixture approaching equilibrium without a catalyst. Which of the graphs below shows the effect of a catalyst on the same reaction mixture?

%P

t

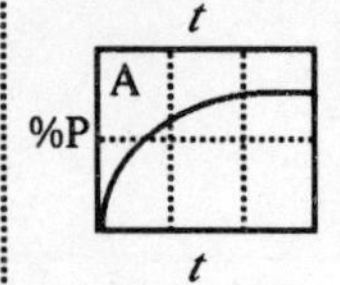

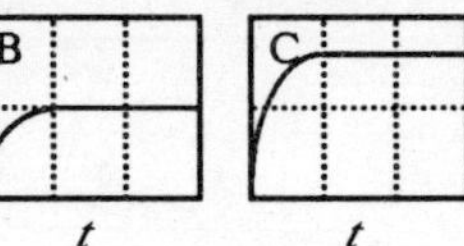

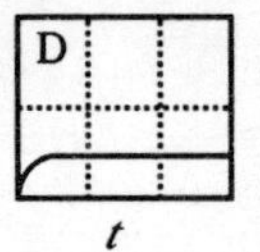

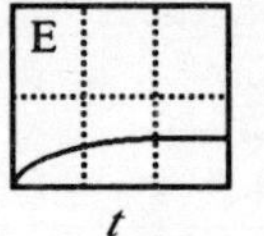

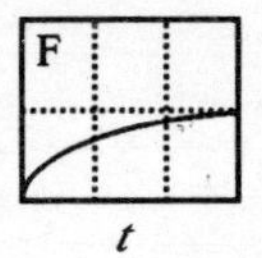

(1)

11. The composition of a hydrogen/oxygen mixture does not change with time. Is the mixture at equilibrium? Describe how you would confirm your answer experimentally. (3)

12. Are the following statements about the effect of catalysts on reversible reactions true or false? A catalyst ...
    (a) increases the forward rate but has no effect on the reverse rate.
    (b) causes more product to be formed at equilibrium.
    (c) establishes the same equilibrium sooner.
    (d) has no effect on an equilibrium.
    (e) increases the forward rate and decreases the reverse rate.
    (f) does not disturb an equilibrium.
    (g) decreases forward and reverse activation energies. (7)

13.

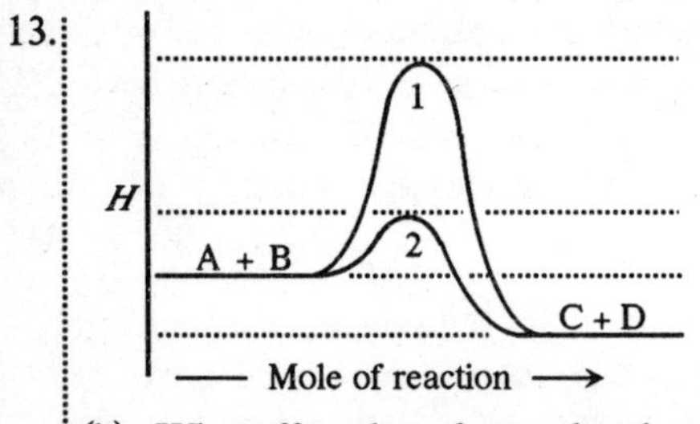

The diagram shows the enthalpy changes for the equilibrium $A(g) + B(g) \rightleftharpoons C(g) + D(g)$, with and without a catalyst.

(a) (i) Which curve, 1 or 2, is obtained with the catalyst? (1)
(ii) Explain your answer to (i). (1)

(b) What effect does the catalyst have on the equilibrium? (2)

(c) (i) What would happen to this equilibrium if temperature were increased? (1)
(ii) Explain your answer to (i). (2)

14. Explain why bubbling hydrogen chloride gas into saturated potassium chloride solution causes the solution to become cloudy. (3)

15. On a cold low pressure day over the Atlantic Ocean there are twenty hydrogen molecules and ten oxygen molecules per 1000 m^3 of air in dynamic equilibrium with the water. Would this number go up or down or stay the same (i) on a hot day, (ii) on a high pressure day? (3)

16. The effect of temperature on nitrogen oxide equilibria is shown below.

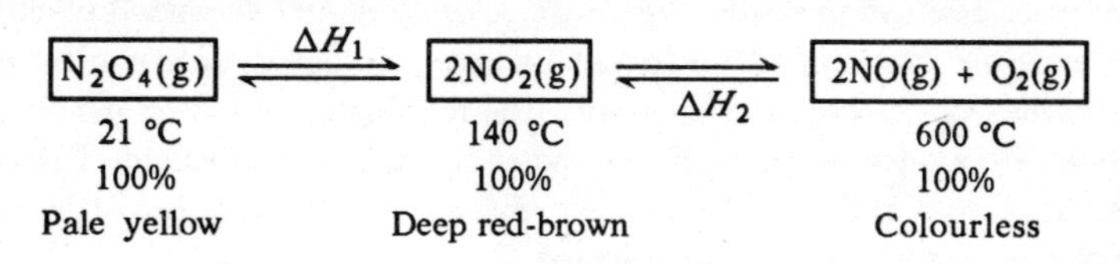

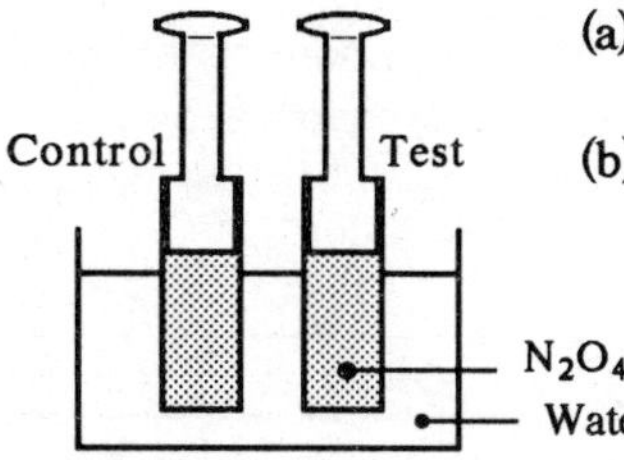

(a) Would the values of (i) ΔH_1 and (ii) ΔH_2 be positive or negative? (2)

(b) If the equilibrium mixture at 50 °C were compressed at constant temperature, what would happen to the colour and why? (3)

17. Iodide ion is oxidised by cerium(IV) ion according to the equation:

$$\underset{\text{yellow}}{Ce^{4+}(aq)} + \underset{\text{colourless}}{I^-(aq)} \rightleftharpoons \underset{\text{colourless}}{Ce^{3+}(aq)} + \underset{\text{brown}}{\tfrac{1}{2}I_2(aq)}$$

The reaction is reversible and proceeds only about half-way to give a mixture containing 25% of each component (starting with equal amounts Ce^{4+} and I^-).

(a) What is the colour of the equilibrium mixture obtained by starting
(i) from Ce^{4+} and I^- (i.e., reactants only); (1)
(ii) from Ce^{3+} and $\tfrac{1}{2}I_2$ (i.e., products only)? (1)

(b) What would happen to the colour if cerium(III) sulphate were added? (1)

(c) In which direction would the equilibrium shift if cerium(IV) sulphate were added? (1)

18. Two routes for the manufacture of epoxyethane (for making antifreeze and detergents) are shown below.

1. $CH_2{=}CH_2 \xrightarrow{Cl_2,\ H_2O} CH_2OH{-}CH_2Cl \xrightarrow{Ca(OH)_2} \underset{\text{Epoxyethane (ethylene oxide)}}{CH_2{-}O{-}CH_2} + CaCl_2$

2. $CH_2{=}CH_2 \xrightarrow[\text{Ag catalyst}]{O_2\ (\text{air})} CH_2{-}O{-}CH_2$

Route 1 has a better yield (70%) than route 2 (60%). Yet, despite this and the hazardous nature of route 2, route 2 is the one used by industry.

(a) Give two reasons why route 1 is not used. (2)

(b) Why is route 2 hazardous? (1)

19. The siting of an industrial plant depends (among other things) on whether it is raw material-oriented or product-oriented. A raw material-oriented plant is located near the source of raw materials because these are more expensive to transport than products, and vice versa. The refining of potash on the North Yorkshire Moors, for example, is raw material-oriented: the weight of crude ore is three times that of the product. State whether the following industries are raw material-oriented or product-oriented.

    (i) Polythene; (ii) Steel; (iii) Baking; (iv) Ammonia and fertilisers; (v) Beer; (vi) Aluminium; (vii) Whisky; (viii) Sulphuric acid (made from sulphur); (ix) Sulphuric acid (made from pyrites ore); (x) Styrene (from benzene and ethene); (xi) Polystyrene; (xii) Oil refining. (12)

20. The composition of a typical synthesis gas used in the Haber process:

$$N_2(g) + 3H_2(g) \rightleftharpoons 2NH_3(g)$$

is shown in the table.

Gas	Volume %
Hydrogen	74.3
Nitrogen	24.7
Methane	0.7
Argon	0.3

Data: Essential Chemical Industry, published by Polytechnic of N. London.

(a) Use the equation to account for the percentage volumes of hydrogen and nitrogen in the table. (1)

(b) State the source of (i) the hydrogen, (ii) the nitrogen and (iii) the argon. (3)

(c) The methane in synthesis gas comes from the methanation reaction:

$$CO(g) + 3H_2(g) \rightleftharpoons CH_4(g) + H_2O(g)$$

(i) What is the purpose of this reaction ? (1)

(ii) Why is it important? (1)

The synthesis gas is passed to a catalyst chamber. From there it goes to a condenser where ammonia (about 15% of the mixture) is removed as a liquid. Unreacted gases are recycled.

(d) (i) Name the catalyst. (1)

(ii) Why is it present as porous pellets? (1)

(e) 15% is much less than the equilibrium percentage. Why then is ammonia withdrawn when this stage is reached? (2)

(f) Draw a flow chart of the process. (2)

Argon accumulates in the system, which therefore has to be purged (when the per cent argon reaches 10%).

(g) (i) Why does argon accumulate? (1)

(ii) What effect does dilution with argon have on the equilibrium system?* (1)

(iii) Explain your answer to (ii). (2)

**In answering this question consider what dilution with argon is equivalent to.*

21. ***Production costs***

These are the costs of (a) raw materials, (b) depreciation, (c) labour, (d) utilities and (e) overheads. Write a brief comment on each of these items. (10)

22.

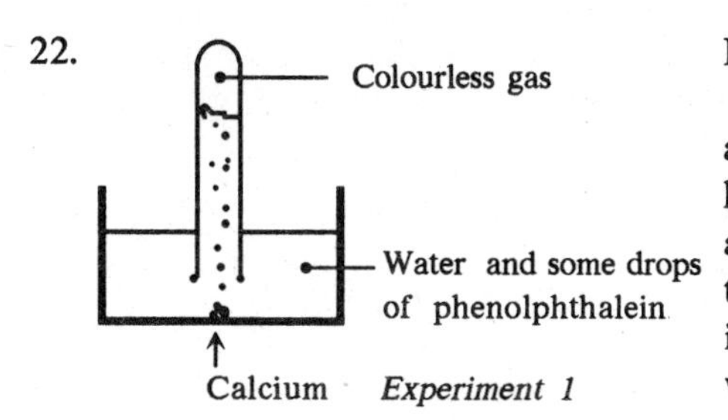

Experiment 1

In an investigation of the equilibrium

$$H^+(aq) + OH^-(aq) \rightleftharpoons H_2O(l)$$

a pupil argued that the removal of hydrogen ion should create alkalinity, and that this could be confirmed with the indicator phenolphthalein, which is red in alkali. Two experiments were tried.

Experiment 1

(a) Why should the removal of hydrogen ion create alkalinity? (2)

(b) In what way does phenolphthalein confirm that this has happened? (1)

(c) (i) Name the colourless gas. (1)

(ii) Describe a test for this gas. (2)

(d) (i) Write an ionic equation for the reaction between calcium and water. (1)

(ii) What kind of reaction is this? (1)

(e) The hydrogen ion concentration is low yet the reaction is fast. Explain. (1)

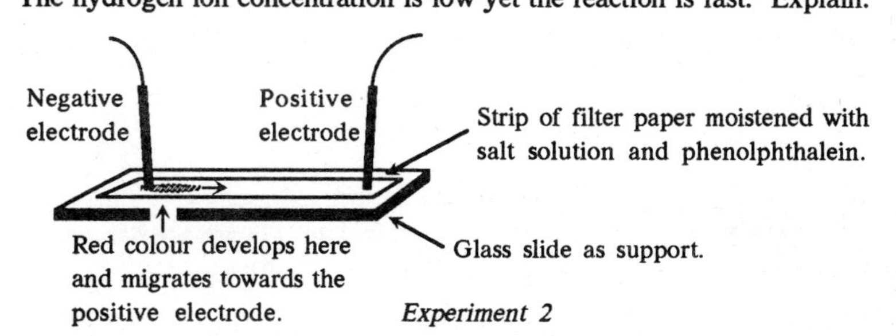

Experiment 2

Experiment 2

(f) (i) Which ion is discharged at the negative electrode? (1)

(ii) Write an ion-electron equation for its discharge. (1)

(g) (i) What effect will this discharge have on the equilibrium? (1)

(ii) State two pieces of evidence from the experiment that confirm this. (2)

23. State the pH of each of the following solutions assuming full ionisation.

(a) Hydrochloric acid, $HCl(aq)$, $c = 0.1$ mol^{-1}. (1)

(b) Nitric acid, $HNO_3(aq)$, $c = 0.0001$ mol L^{-1}. (1)

(c) Sulphuric acid, $H_2SO_4(aq)$, $c = 0.005$ mol L^{-1}. (1)

(d) Hydrobromic acid, $HBr(aq)$, $c = 10$ mol L^{-1}. (1)

24. State the pH of a solution whose $H^+(aq)$ concentration is (i) 10^{-2} mol L^{-1}, (ii) 10^{-7} mol L^{-1}, (iii) 10^{-13} mol L^{-1}, (iv) 1 mol L^{-1}, (v) 10 mol L^{-1}. (5)

25. What is the final pH when
    (a) 10 mL of sodium hydroxide whose pH is 10 is diluted to 1000 mL. (1)
    (b) 10 mL of hydrochloric acid whose pH is 6 is diluted to 1000 mL. (1)

26. What is the pH of a solution whose pH is unaffected by dilution? (1)

27. Ethanol is normally described as neutral. However, it is very slightly ionised in aqueous solution:

$$C_2H_5OH(aq) \rightleftharpoons C_2H_5O^-(aq) + H^+(aq)$$

One mole of ethanol per litre of solution contributes 10^{-9} mole of hydrogen ion. What is the pH of this solution? Explain your answer. (2)

28. Add 90 mL water.
Transfer 10 mL.
Making a tenfold dilution.

A series of tenfold dilutions of hydrochloric acid starting at c = 1 mol L^{-1} were made and pHs determined with a meter. The experiment was repeated for ethanoic acid. Line graphs were plotted as shown below.

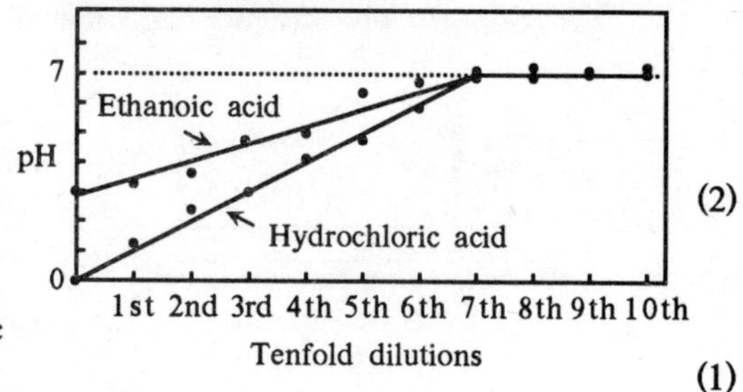

(a) Why are the lines not drawn from point to point? (2)
(b) (i) Using the lines instead of the points, state the pHs of hydrochloric and ethanoic acids at c = 0.1 mol L^{-1}. (1)
(ii) Explain the difference between these pHs. (2)
(c) State the concentration of ethanoic acid after the 5th dilution. (1)
(d) Why is there no further increase in pH after the 7th dilution? (2)

29. State the $[OH^-]$ of a solution in which
    (i) $[H^+] = 10^{-3}$ mol L^{-1}, (ii) $[H^+] = 10^{-9}$ mol L^{-1}. (2)

30. State the $[OH^-]$ of a solution in which
    (i) $[H^+] = 3 \times 10^{-6}$ mol L^{-1}, (ii) $[H^+] = 7.3 \times 10^{-8}$ mol. (2)

31. State the $[H^+]$ of a solution in which
    (i) $[OH^-] = 6 \times 10^{-13}$ mol L^{-1}, (ii) $[OH^-] = 8.7 \times 10^{-2}$ mol L^{-1}. (2)

32. The ionic product of water, $[H^+][OH^-]$, has the value 10^{-14} $(mol\ L^{-1})^2$ at 25 °C (298 K). But the value varies with temperature as shown in the table.

T/K	$[H^+][OH^-]/(mol\ L^{-1})^2$
273	0.1×10^{-14}
280	0.2×10^{-14}
290	0.6×10^{-14}
298	1.0×10^{-14}
310	2.2×10^{-14}
333	10.0×10^{-14}

(a) (i) What happens to the equilibrium $H_2O(l) \rightleftharpoons H^+(aq) + OH^-(aq)$ as temperature increases? (1)
(ii) Is the forward reaction exothermic or endothermic. Explain. (3)
(b) What is the pH of water (i) at 298 K, (ii) at 333 K. (1)
(c) Suggest a reason why the international reference temperature is 25 °C. (1)

33. The percentage ionisations of methanoic acid at increasing dilution are given in the table.

Concentration /mol L^{-1}	Ionisation /%
1	1.4
0.1	4.5
0.01	14
0.001	45

(a) Write an equation for the ionisation of methanoic acid. (1)
(b) What happens to the strength of a weak acid as it becomes more dilute? (1)
(c) Estimate the pH of methanoic acid whose c = 0.001 mol L^{-1}. (1)

34.

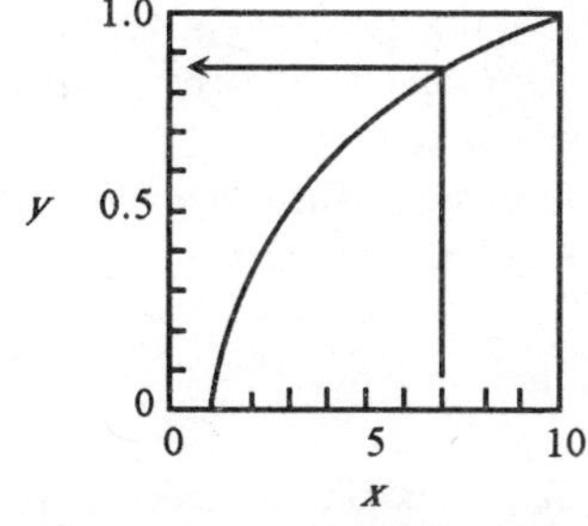

$[H^+]$/mol L^{-1}	*y*	pH
1×10^{-8}		
2×10^{-8}		
5×10^{-8}		
6×10^{-8}		
9×10^{-8}		
10×10^{-8}		

The above graph is used to estimate the pH of a solution from its $[H^+]$. When $[H^+] = x \times 10^{-z}$ mol L^{-1}, pH = $z - y$. If $[H^+] = 7 \times 10^{-4}$ mol L^{-1}, then $x = 7$, $y = 0.86$ and pH = 4 − 0.86 = 3.14.

(a) Copy the table into your note book and use the graph to complete it. (6)
(b) Use the graph to find the pH for $[H^+] = 0.4 \times 10^{-6}$ mol L^{-1}. (1)

35. Given that $[H^+] = 10^{-pH}$, state the $[H^+]$ of a solution whose pH is (i) 0, (ii) 2, (iii) 9, (iv) 14, (v) 11.1, (vi) 5.6, (vii) 2.9, (viii) 0.5, (ix) −1, (x) 15. (8)

36. Why does only one of the four hydrogen atoms in $CH_3COOH(aq)$ ionise? (2)

37. The following questions refer to aqueous solutions of different monoprotic weak acids. State the pH in each case.

(a) c = 1 mol L^{-1}, 0.01% ionised. (1)
(b) c = 0.1 mol L^{-1}, 1% ionised. (1)
(c) c = 0.000 01 mol L^{-1}, 10% ionised. (1)

38.

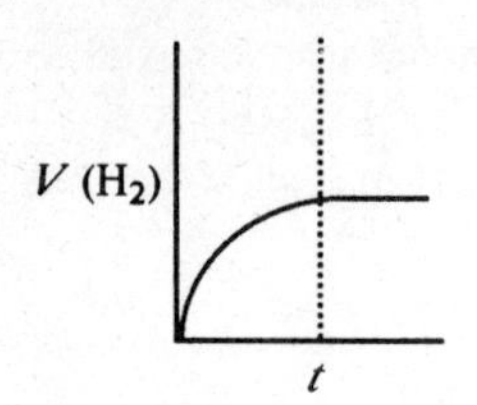

Excess magnesium ribbon was added to 10 mL of 0.1 mol L^{-1} hydrochloric acid. The volume of hydrogen evolved against time is shown in the graph. Copy the graph, alter the lengths of the axes if required, and on it sketch the line that would be obtained with ethanoic acid of the same volume and concentration. (2)

39. Indicators are dyes whose colours are determined by pH. The colours of two indicators are shown below.

pH	0 1 2 3 4	5 6	7 8 9	10 11	12 13 14
Methyl red	Red	c.c.i.	Yellow		
Thymolphthalein	Colourless			c.c.i.	Blue

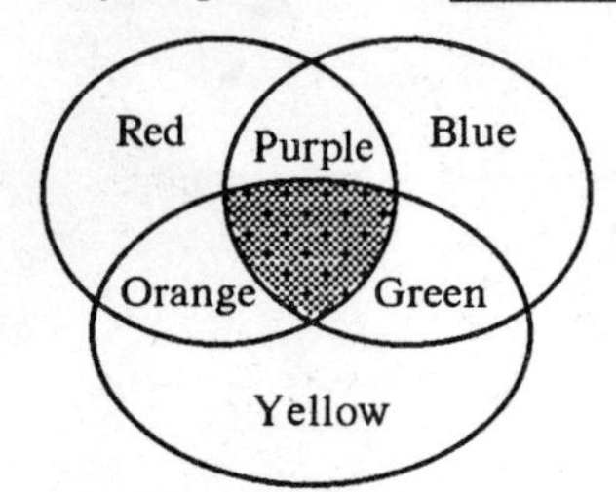

The colour change interval (c.c.i.) is a range of about 2 units over which an indicator is seen to change colour. At the mid point of an interval the colour is a 50% mixture of each: red and blue, for example, give purple.

A mixture of methyl red and thymolphthalein makes a simple universal indicator.

(a) (i) How many colours would this universal indicator exhibit? (1)
(ii) Name the colours and their corresponding pHs. (2)
(b) Why is it wrong to say that yellow is the alkaline colour of methyl red? (1)
(c) (i) Which of the indicators would be suitable for the titration of ethanoic acid with sodium hydroxide? (1)
(ii) Explain your choice in (i). (2)
(d) 1 mL of dilute hydrochloric acid of unknown concentration was added to 10 L of water containing methyl red. Its solution was red. Estimate the pH of the original acid. (1)

40. Identical pieces of excess magnesium were added to 20 mL of hydrochloric and ethanoic acids (c = 0.1 mol L^{-1} for each) containing a little detergent. The foam produced by the reaction in each cylinder was measured at equal time intervals:

(Contd.)

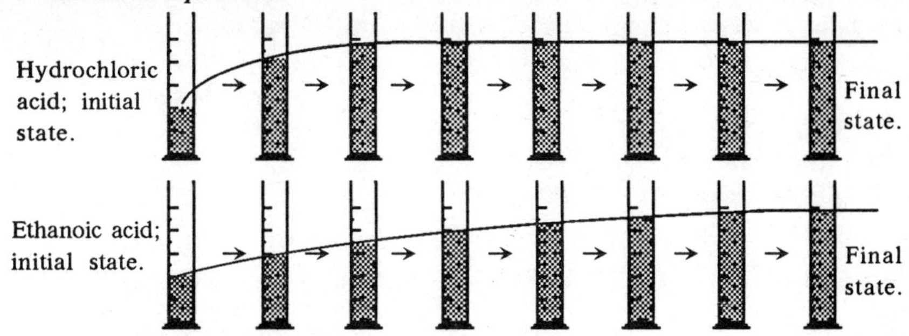

(a) Write an ionic equation for the reaction in each cylinder. (1)

(b) (i) Which is the faster reaction? (1)

(ii) Explain the difference in rate. (2)

(iii) Name two other properties that would support your answer to (ii). (2)

(c) (i) Is there a difference in the final volume of hydrogen? (1)

(ii) Offer a reason for the observation in (i). (2)

41. The percentage ionisations of three halogenated acids ($c = 1 \text{ mol L}^{-1}$) are given in the table.

Acid	Ionisation/%
1. CH_2ICOOH	2.6
2. $CH_2BrCOOH$	3.6
3. $CH_2ClCOOH$	3.7

(a) Name acid 3. (1)

(b) (i) Write an equation for the ionisation of acid 2. (1)

(ii) Explain why ionisation occurs. (2)

(c) Which is the strongest (least weak) of the three acids? (1)

(d) (i) Why do the acids differ in percentage ionisation? (3)

(ii) Where would CH_3COOH fit into the table? (1)

(e) Estimate the pH of acid 1. (1)

42. Carbon dioxide dissolves in water to give an acid solution:

$$H_2O(l) + CO_2(g) \rightleftharpoons 2H^+(aq) + CO_3^{2-}(aq).$$

The forward reaction is exothermic. How would the following changes affect the pH of the solution:

(i) increase in temperature; (1)

(ii) increase in pressure; (1)

(iii) addition of sodium carbonate. (1)

43. Ammonium hydroxide is a weak base. It is approximately 0.001% ionised when its concentration is 0.01 mol L^{-1}.

(a) What is meant by 'weak' base? (1)

(b) Write an equation for the ionisation of ammonium hydroxide. (1)

(c) State (i) its $[OH^-]$ and (ii) its pH. (2)

44. Ammonia dissolves readily in water to give an alkaline solution:

$$NH_3(g) + H_2O(l) \rightleftharpoons NH_4^+(aq) + OH^-(aq).$$

The forward reaction is exothermic. How would the following changes affect the pH of the solution:

(a) (i) an increase in temperature; (1)
(ii) an increase in pressure; (1)
(iii) the dissolving of more ammonia gas. (1)

Methylamine has a similar reaction with water but is not as weak.

(b) Write an equation for the reaction between methylamine and water. (1)
(c) Explain the comment 'but is not as weak'. (1)

45. The first two amines, methylamine and ethylamine, are gases; like ammonia, they dissolve in water to give alkaline solutions:

$$RNH_2(g) + HOH(l) \rightleftharpoons RNH_3^+(aq) + OH^-(aq).$$

To establish whether they are weaker or stronger than ammonia, the pHs of equimolar solutions were determined. The results are shown in the table.

Solution	pH
Ammonium hydroxide	11.6
Methylammonium hydroxide	12.3
Ethylammonium hydroxide	12.4

(a) Write a formula for methylammonium hydroxide. (1)
(b) (i) What does 'equimolar' mean? (1)
(ii) Why must equimolar solutions be used in this experiment? (1)
(c) (i) Which is the strongest of the three bases? (1)
(ii) Explain your answer to (i). (1)

46. Construct a table with suitable headings and classify the following acids and bases as strong or weak: HNO_3, CH_3COOH, KOH, $C_2H_5NH_2$, CH_3NH_2, HI, $Ca(OH)_2$, H_2CO_3, $NaOH$, H_2SO_3, $RbOH$, HCN, NH_3, HBr, H_2SO_4, $HClO_4$, HF, $HCOOH$, HCl, C_2H_5COOH, C_6H_5OH, C_2H_5OH. (11)

47. State the pH of each of the following solutions assuming full dissociation.

(a) Sodium hydroxide, $NaOH(aq)$, $c = 0.1$ mol L^{-1}. (1)
(b) Potassium hydroxide, $KOH(aq)$, $c = 0.001$ mol L^{-1}. (1)
(c) Calcium hydroxide, $Ca(OH)_2(aq)$, $c = 0.000\,05$ mol L^{-1}. (1)

48. A solution of sodium sulphide, Na_2S, is alkaline.

(a) Name and write the formula of the acid that gives this salt. (1)
(b) (i) Is this acid weak or strong? (1)
(ii) Explain why Na_2S gives an alkaline solution. (3)

(Contd.)

(c) Another sodium salt of this acid is not as alkaline as Na_2S.
(i) Write the formula of the other salt. (1)
(ii) What is the common name of a salt like this. (1)
(iii) Explain why it is not as alkaline. (2)

49. A farmer found that the use of ammonium sulphate fertiliser , $(NH_4)_2SO_4$, had made his soil too acid (sour).

(a) Why has the fertiliser had this effect? (2)
(b) What should he add to the soil to neutralise the acidity? (1)

50. The pHs of salt solutions are determined by the nature of the parent acids and bases.

(a) Explain why a solution of ammonium chloride has a pH less than 7. (3)
(b) Explain why a solution of sodium carbonate has a pH greater than 7. (3)
(c) (i) What would be the pH of a solution of potassium nitrate? (1)
(ii) Explain your answer to (i). (3)
(d) Classify the following salts in a table with suitable headings:
KCl, NaBr, $NaHCO_3$, KF, Rb_2SO_4, CH_3COONa, NH_4I, Na_2CO_3, C_2H_5ONa, CH_3NH_3Cl, C_6H_5OK, $CsNO_3$, NH_2CH_2COONa, $CaSO_4$, KCN, $(NH_3CH_2COOH)Cl$, K_2SO_3, KBr, HCOONa, BaF_2. (10)

51. Explain why the addition of ammonium chloride, NH_4Cl, to aqueous ammonia lowers the pH. (3)

52. Explain why the addition of sodium ethanoate, CH_3COONa, to aqueous ethanoic acid raises the pH. (3)

53. Phenol red is an acid/base indicator. It itself is a weak acid and ionises as shown:

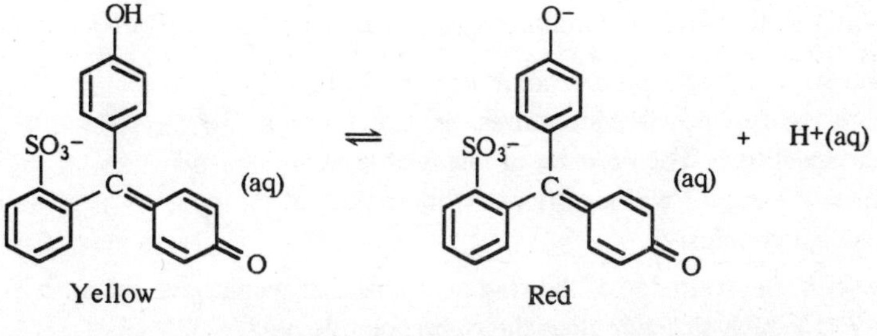

OH, SO_3^-, C, O (aq) $\rightleftharpoons$ O^-, SO_3^-, C, O (aq) + $H^+(aq)$

Yellow — Red

(a) At pH 7.9 phenol red is 50% ionised. What is its colour at this pH? (1)
(b) What effect will added sodium hydroxide have on this equilibrium? (1)
(c) What will be the 'alkaline colour' of this indicator? (1)

UNIT 8
RADIOISOTOPES

1. Zinc is a mixture of five isotopes. Data for these are given in the table. Use these data to calculate the relative atomic mass (A_r) of zinc.

Isotope	Relative isotopic mass	Abundance/%
Zn-64	63.9	48.9
Zn-66	65.9	27.8
Zn-67	66.9	4.1
Zn-68	67.9	18.6
Zn-70	69.9	0.6

(3)

2. 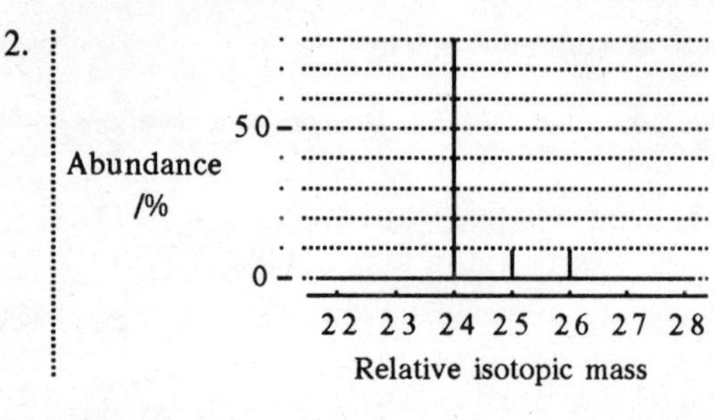

A simplified mass spectrum for magnesium is shown opposite.

(a) (i) How many isotopes does magnesium have? (1)

(ii) Which isotope is the most abundant? (1)

(b) Calculate the relative atomic mass of magnesium. (3)

3. Boron consists of two isotopes, B-10 and B-11. The relative atomic masses of these isotopes are 10.01 and 11.01 respectively. The relative atomic mass of the element as a whole is 10.81. Calculate the abundance of each isotope. (2)

4. Copper is a mixture of two isotopes whose abundances are 69% and 31%. A_r(copper) is 63.5. One of the isotopes has a relative mass of 62.9.

(a) Is the relative mass of the other isotope smaller or greater than 62.9? (1)

(b) Explain your answer to (a). (1)

(c) Calculate the relative mass of the other isotope. (2)

5. Write molecular formulae for and state the molecular masses of

(i) carbon dioxide that contains C-14,

(ii) water that contains one H-3 (tritium) atom per molecule,

(iii) methanol that contains oxygen with a mass number of 17. (3)

6. Fragmentation of a pure organic liquid, C_2H_6O, which has two possible structures, gave a complex mass spectrum. The *absence* of peaks at relative masses 29 and 17 was significant while a peak at 31 was inconclusive.

Structure 1, CH_3OCH_3. Structure 2, CH_3CH_2OH.

(a) (i) Write formulae of the fragment ions that would give peaks at 29 and 17. (2)

(ii) Which structure does their absence rule out? (1)

(iii) Explain your answer to (ii). (1)

(b) Why is a peak at relative mass 31 inconclusive? (2)

7. A simplified mass spectrum for bromine is shown below.

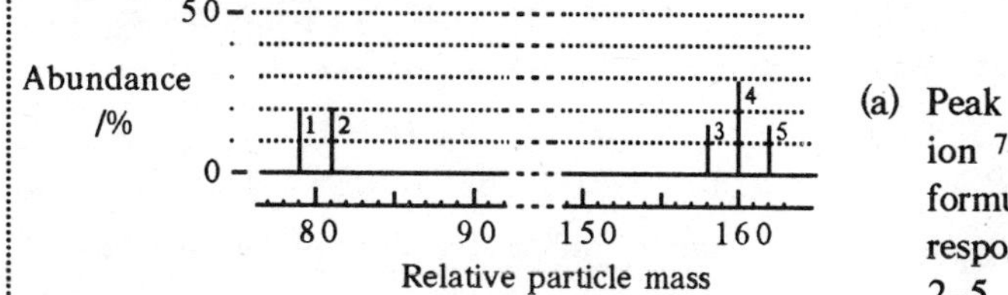

(a) Peak 1 is due to the ion $^{79}Br^+$. Write formulae for the ions responsible for peaks 2–5. (2)

(b) (i) From data in the spectrum state the average M_r of bromine. (1)
(ii) Explain your reasoning in (i). (1)

8. A simplified mass spectrum for chlorine is shown below.

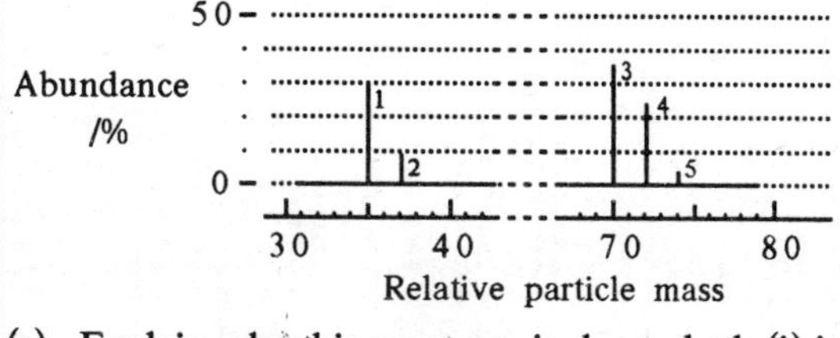

(a) Peak 3 is due to the ion $^{35}Cl_2^+$. Write formulae for the ions causing 1, 2, 4 and 5. (2)

(b) Calculate A_r(Cl). Show your working. (3)

(c) Explain why this spectrum is due to both (i) isotopic variation and (ii) molecular fragmentation. (2)

(d) A third factor, multiple charge, may affect some spectra. At what relative particle mass would a peak for the ion $^{35}Cl^{2+}$ appear? (1)

9. Carbon has a low A.N. and two stable isotopes: C-12 and C-13. Bismuth has the highest A.N. of the stable elements. It has one stable nuclide: Bi-209.

(a) (i) Calculate the average n/p ratio for carbon. (1)
(ii) Calculate the n/p ratio for bismuth. (1)

(b) What happens to the value of the n/p ratio of stable elements as atomic number increases? (1)

10. Write nuclear equations for the following.

(a) α decay of (i) Th-232; (ii) U-238. (2)

(b) β decay of (i) C-14 ; (ii) H-3. (2)

(c) e^- capture by (i) K-40; (ii) Mn-54. (2)

(d) n capture by (i) U-238; (ii) Ar-40. (2)

11. A slow-moving neutron reacts with the N-14 nucleus to give either (i) a C-14 nucleus and a proton, or (ii) a B-11 nucleus and an α particle. A fast-moving neutron gives (iii) a C-12 nucleus and a tritium (H-3) nucleus.

(a) Name the background radiation that initiates these changes. (1)

(b) Write nuclear equations for (i), (ii) and (iii). (3)

12. Solve the nuclear equations.

(a) $^{27}_{13}Al + ^{1}_{0}n \rightarrow ^{4}_{2}He + ^{A}_{Z}X$.

(b) $^{39}_{19}K + ^{1}_{0}n \rightarrow 2\,^{A}_{Z}X + ^{38}_{19}K$.

(c) $^{9}_{4}Be + \gamma \rightarrow ^{8}_{4}Be + ^{A}_{Z}X$.

(d) $^{236}_{92}U \rightarrow ^{85}_{34}Se + ^{A}_{Z}X + 3\,^{1}_{0}n$.

(e) $^{2}_{1}H + ^{A}_{Z}X \rightarrow ^{3}_{2}H + ^{1}_{0}n$.

(f) $^{14}_{7}N + ^{4}_{2}He \rightarrow ^{A}_{Z}X + ^{1}_{1}H$.

(g) $^{1}_{0}n \rightarrow ^{A}_{Z}X + ^{0}_{-1}e$.

(h) $^{84}_{37}Rb \rightarrow ^{84}_{38}Sr + ^{A}_{Z}X$. (8)

13. The following sequence is taken from the radioactive series in the data book.

$$^{234}_{91}P \xrightarrow{-X} ^{234}_{92}Q \xrightarrow{-Z} ^{230}_{90}R$$

(a) (i) Identify P, Q and R. (1)
(ii) State which is the most stable. (1)

(b) Name particles x and z. (1)

14. The diagram illustrates the decay of 64 nuclei.

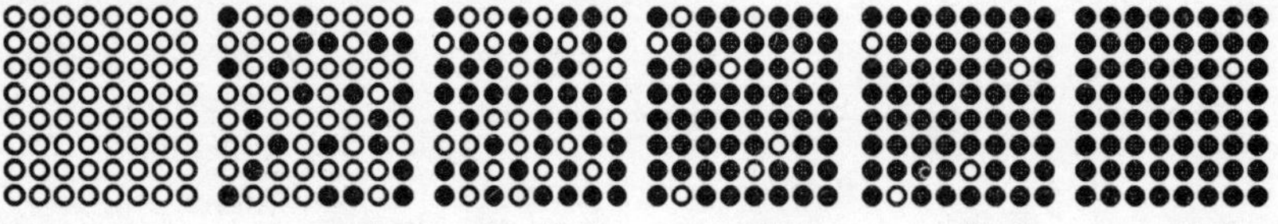

t = 0 | t = 5 d | t = 12 d | t = 24 d | t = 33 d | t = 45 d

(a) Determine $t_{1/2}$ for this nuclide. (3)

(b) If 64 fresh nuclei were selected and the experiment repeated:
(i) How many nuclei would decay by t = 5 d? (1)
(ii) In what way would the 'box' at t = 5 d now be different? (1)
(iii) Why would it be different? (1)

15. What fraction of a radioisotope remains after (i) one half-life, (ii) two half-lives, (iii) three half-lives, (iv) four half-lives. (2)

16. The fraction of a radioisotope remaining after n half-lives is $(^1/_2)^n$. Use this relation to calculate the fraction remaining after (i) $^1/_2\ t_{1/2}$, (ii) 50 $t_{1/2}$. (2)

17. State whether the following statements are true or false.

(a) 1 g of K-40 has the same half-life as 1 kg of K-40. (1)

(b) Radium iodide (RaI_2) has a greater half-life than radium fluoride (RaF_2). (1)

(c) Pb-214 has the same half-life as Pb-212. (1)

(d) 2 moles of H-3 have the same half-life as 1 mole of H-3. (1)

(e) Hot radon has the same half-life as cold radon. (1)

18. Radiocarbon dating of limestone (calcium carbonate) deposits at the bottom of a shallow lake gave the age of the limestone as 10 000 years. Fossil dating gave the limestone's age as 3000 000 years.

(a) What is the reason for this contradiction? (2)

(b) Why would the error not arise with material like leather and wood? (1)

19.

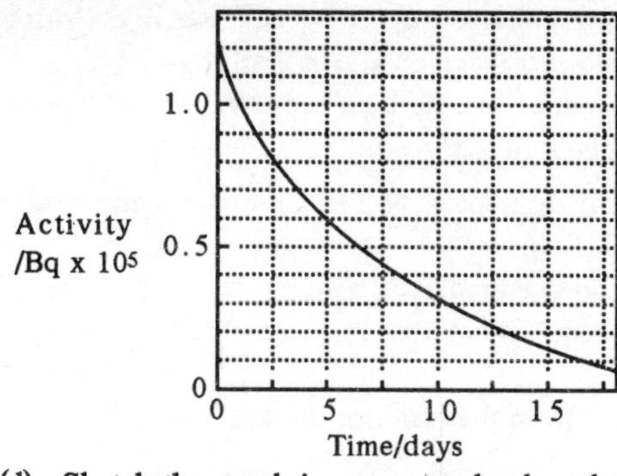

The decay curve for a radioisotope is shown opposite.

(a) State the half-life of this isotope. (1)

(b) Identify the isotope from the decay series in the data book. (1)

(c) Write nuclear equations for the decay of this and its daughter isotope that leads to a stable end product. (2)

(d) Sketch the graph in your note book and draw a curve for the decay of

(i) half the mass of the same isotope; (2)

(ii) an isotope with the same initial activity but half the half-life period. (2)

20. The initial quantity of a radioisotope and the period over which it decays is stated below. How much remains after this period?

(i) 8 mol over 2 $t_{1/2}$; (ii) 12 g over 3 $t_{1/2}$; (iii) 10 L over 4 $t_{1/2}$;

(iv) A quantity whose initial activity was 3×10^8 Bq over 2 $t_{1/2}$. (4)

21. The final quantity of a radioisotope and the period over which decay has occurred is stated below. What was the initial quantity?

(i) $\frac{1}{4}$ mol after 2 $t_{1/2}$; (ii) 10 g after 1 $t_{1/2}$; (iii) 250 mL after 4 $t_{1/2}$;

(iv) A quantity whose final activity is 5×10^{10} Bq after 3 $t_{1/2}$. (4)

22. The initial and final quantities of a radioisotope are given. State how many half-lives have elapsed in each case.

(i) 1 mol → $\frac{1}{4}$ mol; (ii) 10 g → $1\frac{1}{4}$ g; (iii) 1 L → 62. 5 mL;

(iv) Quantities whose activities change: 1.6×10^7 Bq → 8×10^6 Bq. (4)

23. Three kinds of radiation are emitted by radioactive materials as shown in the table. Their ability to penetrate matter varies as shown in the diagram below.

Name	Alpha (α)	Beta (β)	Gamma (γ)
Mass/u	(i)	1/2000	0
Charge	2+	(ii)	(iii)
Symbol	(iv)	${}^{0}_{-1}e$	γ

α source

β source

γ source

Paper Al Pb

(a) Complete (i) to (iv) in the table. (2)

(b) Radiation is used to destroy cancer cells. Consider the information in the diagram and state which radiation should be used to treat (i) the skin, (ii) tissue just under the skin and (iii) deeper tissues. (2)

24. A dilute salt solution containing a minute quantity of $^{24}NaCl$ was injected into the bloodstream of a patient who had just received a skin graft. Na-24 is radioactive and emits γ rays, which can be located by a detector. A whole body scan showed low-level γ activity except over the graft.
    (a) Na-24 is made by neutron bombardment of N-23. Give two reasons why neutrons are used to create radioisotopes. (2)
    (b) (i) What is the name of the technique employed here. (1)
    (ii) The salt is said to be 'labelled'. What does this mean? (1)
    (c) What are γ rays? (1)
    (d) What conclusion must be drawn from the result of the scan? (1)

25. When a U-235 nucleus absorbs a neutron it splits in two and releases two or three neutrons and energy. Boron rods are inserted to absorb neutrons and control the chain reaction. The diagrams below illustrate chain reactions.

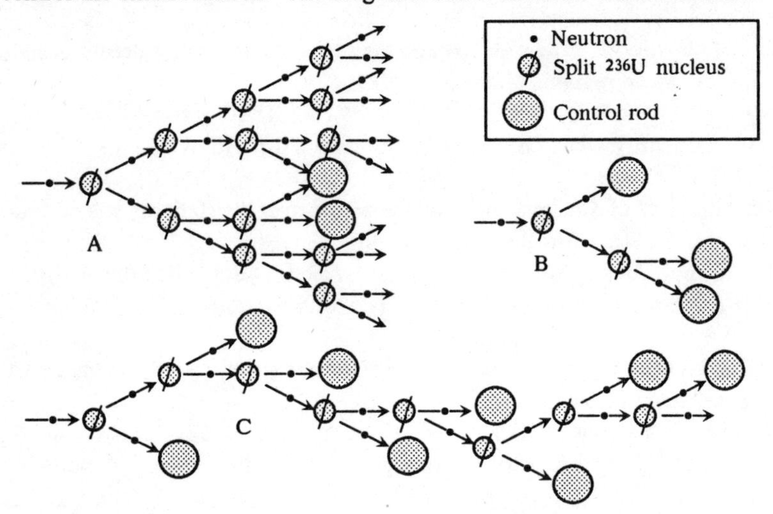

    (a) Identify the reaction that (i) is not sustained, (ii) is occurring at a steady rate, (iii) is increasing. (2)
    (b) Which sequence is used (i) in nuclear power stations, (ii) explosive devices? (1)

26. Several pairs of woven rope sandals unearthed in Fort Rock Cave (Oregon) had a C-14 activity approximately $1/4$ that of fresh material. Estimate the age of the sandals given that $t_{1/2}$(C-14) is 5700 years. (1)

27. C-14 is unstable and decays by β loss. 15.3 β particles are ejected per minute per gram of new wood. Wood from a house in the ancient Lebanese city of Byblos gave a count of 8.1 β particles min^{-1} g^{-1}. Determine the age of the wood given that $t_{1/2}$(C-14) is 5700 years. (3)

28. The Fe-56 nucleus and its neighbouring nuclei in the Periodic Table represent a high point in stability and a low point in energy: the transmutation of lighter *and* heavier nuclei into these 'middle nuclei' is exoenergetic.

Nuclei of small A.N.	Reaction A → Energy lost	'Middle nuclei'	← Reaction B Energy lost	Nuclei of large A.N.

(a) Why are some nuclei more stable than others? (1)
(b) Explain the meaning of 'transmutation'. (1)
(c) Name reactions A and B. (1)
(d) Research into reaction A as a source of energy has met with limited success. What is the main reason for this? (1)
(e) Reaction B, in contrast, has been used successfully as a source of energy.
(i) Name one isotope used in this reaction. (1)
(ii) Describe very briefly how the energy is released from this fuel. (2)
(ii) Write a possible nuclear equation for the reaction. (1)

29. $E = mc^2$ states the relationship between mass and energy. E is expressed in J, m in kg; c is the speed of light, 3×10^8 m s^{-1}.

(a) Imagine that two deuterium nuclei fuse to give one helium atom:

$$2\,^{2}_{1}\text{H} \longrightarrow \,^{4}_{2}\text{He} \quad \ldots 1$$

Mass (m): 4.0282×10^{-3} kg $\qquad$ 4.0026×10^{-3} kg

(i) State the mass change, Δm, in kg. (1)
(ii) Use $E = mc^2$ to express Δm as an energy change, ΔE, in J mol^{-1}. Note that J = kg x (m s^{-1})^2. (2)
(iii) If deuterium could be made to supply us with fusion energy, it would outlast our finite fuels by 50 billion years! Explain. (2)
(iv) Fusion of deuterium and tritium (H-3) to give He-4 has been achieved on a small scale. Write an equation for this reaction. (1)

(b) Two mole of (diatomic) hydrogen burn in oxygen:

$$2H_2 + O_2 \longrightarrow 2H_2O \quad \ldots 2$$

Refer to the Data Book and state the heat change, ΔH, for this reaction. (1)

(c) (i) Which reaction is the better source of energy? (1)
(ii) By what factor is it a better source? (1)
(iii) Why is the comparison between reactions 1 and 2 valid? (1)

30. A problem with nuclear reactors is the disposal of waste: some fission products have long half-lives. ^{90}Sr and ^{137}I each have half-lives of about 28 years. In general, radioisotopes are not considered safe until 20 half-lives have elapsed.

(a) For how long would waste containing these isotopes have to be stored? (1)
(b) What fraction of the original activity remains after 20 half-lives? (1)

ANSWERS

Unit 1

1. (a)

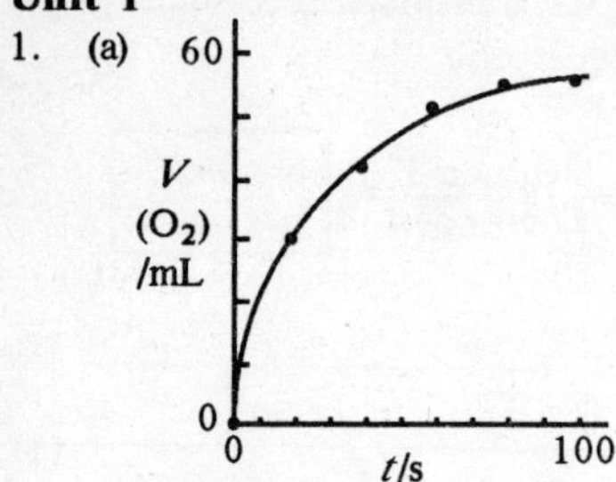

(b) 55 s. (c) Average rate $= \dfrac{V_2 - V_1}{t_2 - t_1}$

(i) $= \dfrac{38 - 30}{30 - 20} \dfrac{\text{mL}}{\text{s}}$

$= 0.8 \text{ mL s}^{-1}$

(ii) $= \dfrac{47 - 42}{50 - 40} \dfrac{\text{mL}}{\text{s}}$

$= 0.5 \text{ mL s}^{-1}$

(d) The rate of reaction at c(ii) is smaller because the concentration of hydrogen peroxide is smaller.

2.

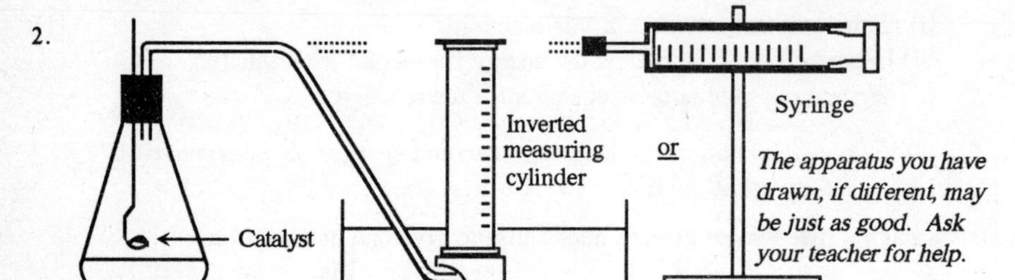

3. (a) (i) Reactant.
   (ii) It decreases.
   (b) mol L^{-1} min^{-1}.

(c)

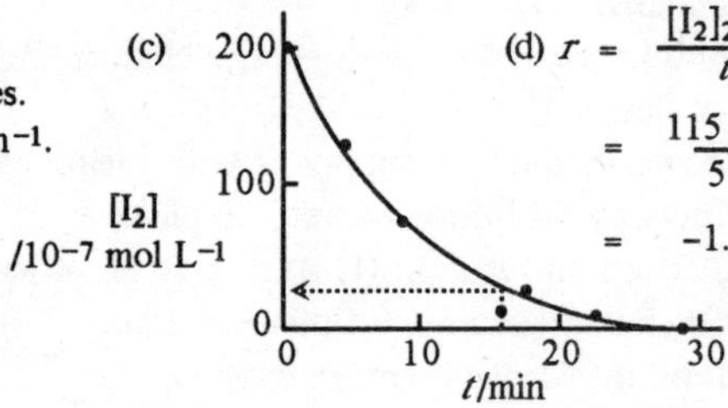

[I_2] /10^{-7} mol L^{-1} (axis: 0, 100, 200); t/min (axis: 0, 10, 20, 30)

(d) $r = \dfrac{[I_2]_2 - [I_2]_1}{t_2 - t_1}$

$= \dfrac{115 - 200}{5 - 0} \dfrac{10^{-7} \text{ mol L}^{-1}}{\text{min}}$

$= -1.7 \times 10^{-6} \text{ mol L}^{-1} \text{ min}^{-1}$

(e) The concentration at 16 min; it should have been about 3.0×10^{-6} mol L^{-1}.

4. (a) A variable is a physical quantity that can be altered during a reaction.
   (b) Temperature, pressure, concentration, volume, mass, ...
5. A catalyst or enzyme.
6. (a) (i) Time; (ii) Mass.
   (b) The independent (or controlled) variable is the one we select at intervals. The dependent variable is the one whose value is determined by the value of the independent variable. (The independent variable is plotted on the x axis, the dependent on the y axis.)

(c) The rate of reaction at 3 min is found by calculating the average rate between 2 and 4 min. This is correct when the curve is circular.

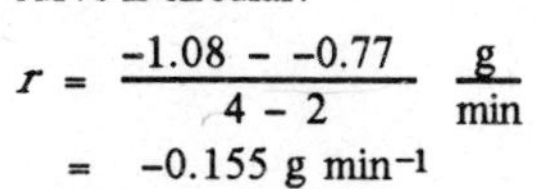

$$r = \frac{-1.08 - -0.77}{4 - 2} \frac{\text{g}}{\text{min}}$$

$= -0.155$ g min^{-1}

(The instantaneous rate at 3 min is also the slope of tangent AB.

$r = -0.8/5.25$ g min^{-1}

$= -0.152$ g min^{-1})

Time/min
0 1 2 3 4 5 6 7 8
Mass /g
0
A
−1.0
B

(d) $CaCO_3 + 2HCl \rightarrow CaCl_2 + H_2O + CO_2$

(e) 1 mol ... 1 mol

$m = 100$ g ... $m = 44$ g

44 g CO_2 is obtained from 100 g $CaCO_3$

1.3 g CO_2 (from graph) is obtained from (100/44) x 1.3 g = 2.95 g

7. (a) $r \propto 1/t$.

(b) r for 1 mol L^{-1} acid is 0.018 s^{-1}, for 2 mol L^{-1} acid, 0.067 s^{-1}, and for 3 mol L^{-1} acid, 0.143 s^{-1}.

(c)

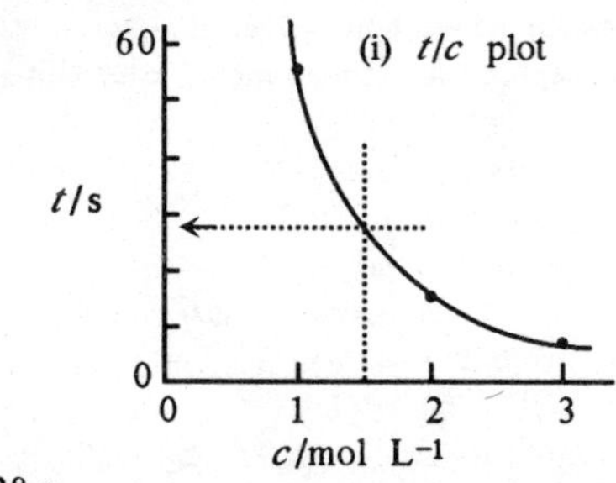

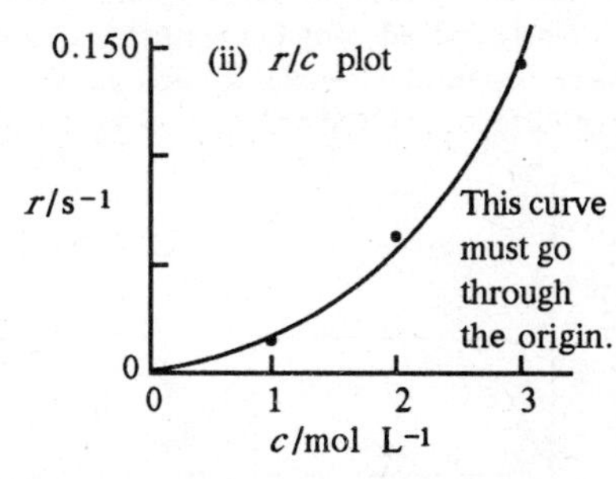

(d) 28 s.

8. (i) Rusting; (ii) Neutralisation. Many other answers.

9. (a) The greater the concentration, the greater the rate.

(b) The second reaction: doubling [B] doubles r, and so on.

10. (a) Experiment 1. (b) Curve 1 has the steepest initial slope. (c) Experiment 1.

(d) Experiment 1 has most thiosulphate to react away and so it is both fastest to begin with and longest to reach completion.

(e) For experiment 1, 20 mL 0.1 mol L^{-1} is diluted to 200 mL total volume. Hence c_1 is 0.01 mol L^{-1}. c_2 and c_3 are 0.0075 and 0.005 mol L^{-1} respectively. (Use $V_1c_1 = V_2c_2$ to solve this kind of problem.)
t_1 and t_3 are 2 and 12 s resp..
r_1 and r_2 are 0.5 and 0.2 s^{-1} resp..

(f)

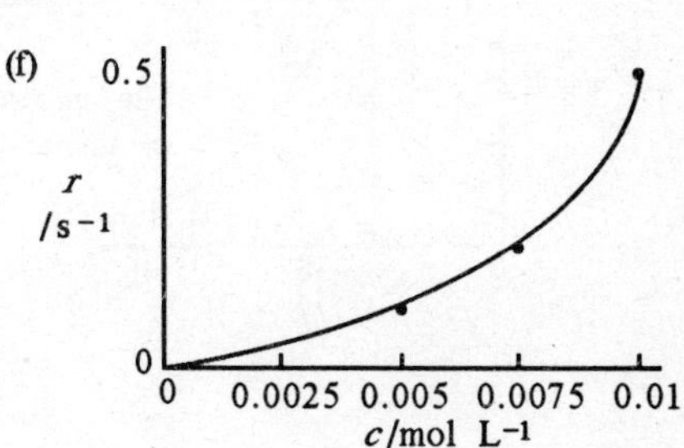

11. (a) Concentration (of sodium thiosulphate).
    (b) (i) Volume of reaction mixture; (ii) The sum 'volume of 0.1 mol L^{-1} thiosulphate + water' is constant.
12. (a) D. (b) A. (c) B. (d) C.
13. (a) Six (H_2/H_2, H_2/I_2, H_2/HI, I_2/I_2, I_2/HI and HI/HI).
    (b) (i) HI/HI collisions; (ii) H–I + H–I → H···I / H···I (dotted bonds between H and H, and between I and I) → H–H + I–I
    (c) Collisions should be energetic and have correct orientation.

14.

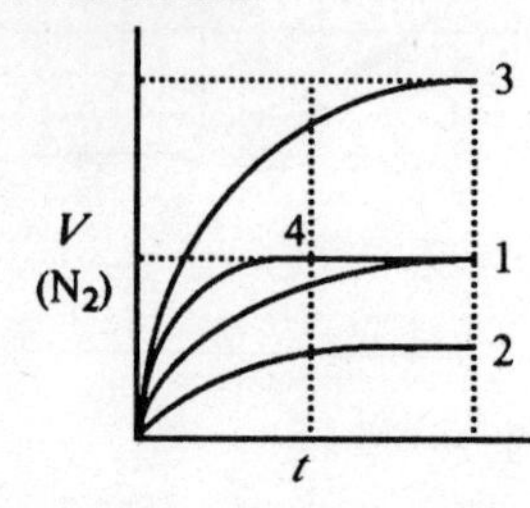

Comments
2 is like 1 after 1 is half-way to completion. So 2 finishes before 1 and $V(N_2)$ is halved.
3 is like two 1s running in parallel. $V(N_2)$ is doubled and is delivered in the same time.
4 is faster than 1. $V(N_2)$ is the same but is delivered in less time.

15. Particle size. The finely divided graphite, which is now exposed after the loss of the iron, has a very large surface area and reacts rapidly with the air.
16. (a) Finely divided. (b) Large surface area allows rapid reaction even at r.t..
17. The carbon disulphide evaporates leaving the phosphorus in a finely spread state with a large surface area. $2P + 2\frac{1}{2}O_2 \rightarrow P_2O_5$ or $P_4 + 5O_2 \rightarrow 2P_2O_5$.

18.

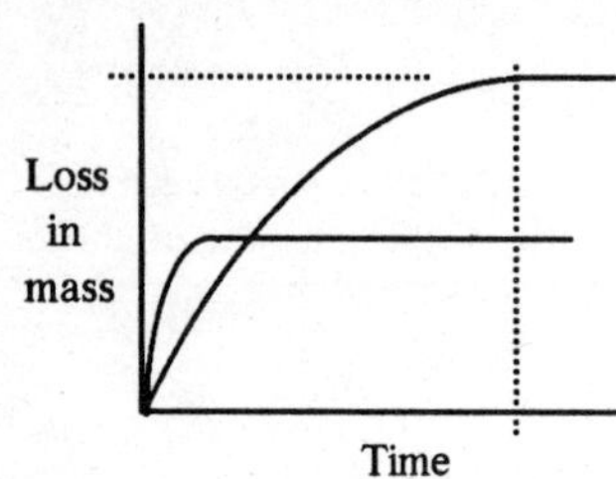

19. (a) 6 cm^2.
    (b) (i) 1 nm^3 = $(10^{-7}\ cm)^3$
    = 10^{-21} cm^3.
    (ii) $(1\ cm^3)/(10^{-21}\ cm^3) = 10^{21}$.
    (c) (i) 6 nm^2 = 6 x $(10^{-7}\ cm)^2$
    = 6 x 10^{-14} cm^2.
    (ii) 10^{21} x 6 x 10^{-14} cm^2
    = 6 x 10^7 cm^2.
    (d) Surface area: the surface area of a cm cube broken down to a 'nanometre powder' increases from 6 cm^2 to 6 x 10^7 cm^2, a million-fold increase!

20.

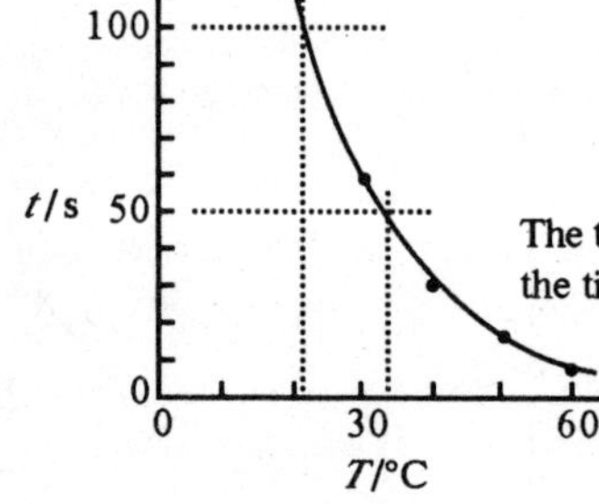

The temperature rise that doubles the rate (halves the time) is approximately (34 – 21) °C = 13 °C.

21. (a) Exothermic (because ΔH has a negative value).
(b) High activation energy.
(c) Catalase.
(d)

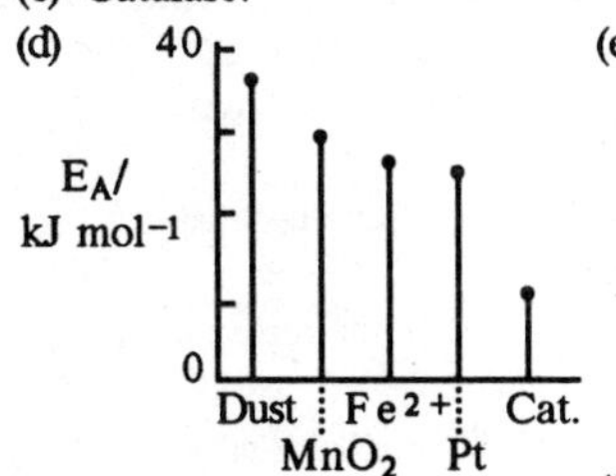

(e)

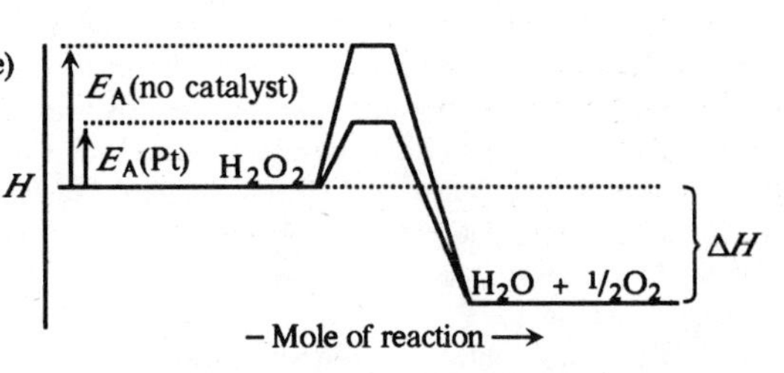

(f) Heat produced and replacement of reactant chemicals.

22. (a) The total number of molecules.
(b) and (c)

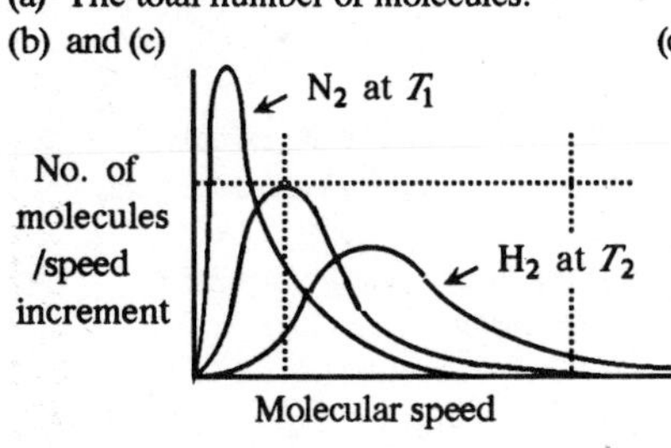

(d) Average speed of H_2 molecules at T_2 is greater than at T_1. So peak moves to the right and becomes lower because area under curve is constant.
Temperature is determined by $\frac{1}{2}mv^2$. N_2 molecules are heavier than H_2 molecules and so their average speed at T_1 is smaller. Peak shifts to the left and increases in height to conserve area.

23. (a) and (b)(i)

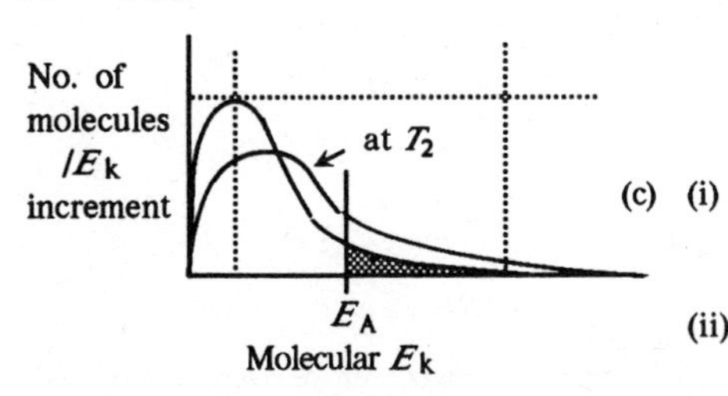

(ii) In this context, activation energy, E_A, is the minimum molecular kinetic energy for reaction to occur on collision. (E_A is scaled up to a value per mole.)
(c) (i) The peak moves to the right and becomes lower because area is conserved.
(ii) The area under the curve to the right of E_A increases rapidly.
(d) (i) No effect; (ii) The value of E_A decreases (moves to the left); (iii) Area becomes greater because E_A moves to the left. (Note how the 'reactive' area under the curve can increase for two quite separate reasons: either the curve slews to the right at higher temperature or E_A moves to the left when a catalyst is introduced.)

24. (a) Greater surface area.
(b) The platinum is involved in the reaction (as the activated complex). But involvement is temporary and the regenerated platinum is not deposited in the same place.
(c)

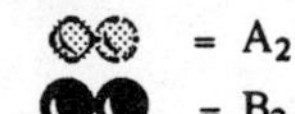

25. (a) E. Zinc, being divalent like magnesium, gives the same volume of hydrogen but the reaction is slower because zinc is less reactive.
    (b) B. Twice the amount of magnesium liberates twice the volume of hydrogen. The reaction takes just the same time, however, provided the acid is in large excess.
    (c) F. The reaction is very fast because sodium is very reactive. But only half the volume of hydrogen is obtained because sodium is monovalent.
    (d) E. The same volume of hydrogen is obtained but the reaction is slower because turnings have a smaller surface area per mole than powder.
    (e) A. The same volume of hydrogen is obtained but the reaction is faster because of the higher temperature.
    (f) E. The same volume of hydrogen is obtained but the reaction is slower because the acid is less concentrated.
26. (a) (i) $\Delta H = H_2 - H_3$; (ii) $\Delta H = H_7 - H_8$; (iii) $E_A = H_1 - H_3$; (iv) $E_A = H_5 - H_7$.
    (b) (i) Positive; (ii) Endothermic.
    (c) (i) $E_A = H_1 - H_2$; (ii) $H_5 - H_8$.
    (d) E_A is the potential energy of the activated complex minus that of the reactants (forward reaction) or minus that of the products (reverse reaction). Values are quoted per mole of reaction.
27. (a) (i) Potential energy; (ii) If temperature is constant, kinetic energy is constant. Energy lost or gained must therefore be potential in origin.
    (b) 25 °C (298 K). (c) Exothermic.
28. (a)

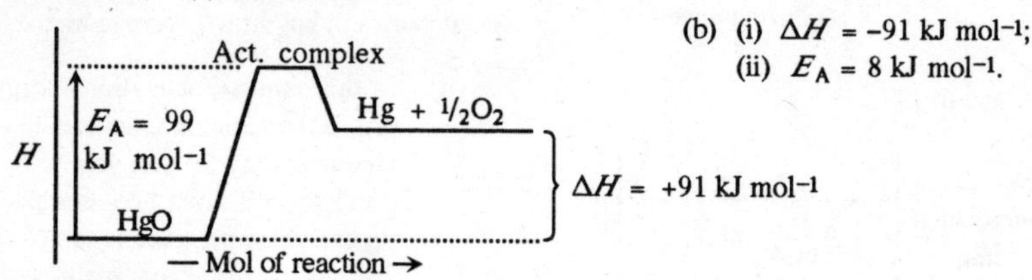

(b) (i) $\Delta H = -91\ \text{kJ mol}^{-1}$;
(ii) $E_A = 8\ \text{kJ mol}^{-1}$.

29. (i) $\Delta H = 2 \times +91\ \text{kJ mol}^{-1} = +182\ \text{kJ mol}^{-1}$; (ii) $E_A = 198\ \text{kJ mol}^{-1}$.
30. (a) $E_A = 330\ \text{kJ mol}^{-1}$.
    (b) (i) E_A becomes smaller; (ii) ΔH is unaffected.
31. The activated complex is an unstable particle formed at the maximum of the potential energy barrier for a reaction. (The complex is described as short-lived or transient.)
32. (a) False: can be solid, liquid, gas or solution of same. (b) False: is chemically unchanged but may be physically changed. (c) False: but preferable for a solid catalyst. (d) True. (e) False. (f) False: is very much involved but is left over afterwards. (g) True. (So-called "negative catalysts" are really poisons and inhibitors that destroy catalysts and enzymes.) (h) False: light is a form of energy and may initiate and speed up reactions just as would heat.
33. (a) The reaction using catalase. (b) The reaction using catalase. (c) (i) Increase speed; (ii) No effect. (d) $123\ \text{kJ mol}^{-1}$. (e) Heterogeneous: different phases involved. (A phase is a solid, liquid, gas or solution.)

Unit 1 Answers

34. A: homogeneous. B: heterogeneous. C: colloidal (enzymic). D: heterogeneous.
35. (a) (i) Poisons; (ii) Poisons are adsorbed at the 'active sites' on the catalyst surface and so prevent adsorption of reactants.
    (b) (i) Inhibitors; (ii) Inhibitors fit into the active sites on the enzyme molecules and so block the entry of the substrate molecules.
36. (a) Catalyst.
    (b) (i)
    $$2Al + 3I_2 \rightarrow 2AlI_3$$
    $$2AlI_3 + 1\tfrac{1}{2}O_2 \rightarrow Al_2O_3 + 3I_2$$
    (iii)
    $$2Al + 1\tfrac{1}{2}O_2 \rightarrow Al_2O_3$$
    (ii) Aluminium reacts rapidly with iodine, and aluminium iodide reacts rapidly with oxygen. Iodine is involved in the reaction, speeds it up, and is left over afterwards.
    (c) Electric spark, flame, light.
37. (a) Because ions are involved and these react almost instantaneously. (Reactions between covalently bonded molecules are slow because covalent bonds are slow to break and remake.)
    (b) $Ba^{2+}(aq) + SO_4^{2-}(aq) \rightarrow BaSO_4(s)$
38. In answering this question, remember that E_As always have positive values, whether forward or reverse. Remember also that ΔH for an endothermic reaction ($E_{A,f} > E_{A,r}$) has a positive value. Therefore, $E_{A,f} - E_{A,r} = \Delta H$. The same relation is obtained if you reason it from an exothermic reaction for which $E_{A,f} < E_{A,r}$ and ΔH has a negative value. If you still have trouble with this, try working it out from an enthalpy diagram.
39. (a) The activation energy is increased because it reverts to the uncatalysed value (and the reaction slows or stops).
    (b) No effect.

Unit 2

1. (a) A naturally occurring material (e.g., crude oil) used as a source either of energy or of other materials.
   (b) A useful naturally occurring material that will eventually be used up (e.g., coal).
   (c) A chemical from which other chemicals can be made (e.g., syngas).
   (d) A material used to make saleable goods (e.g., a plastic) or a material sold as it is (e.g., petrol).
   (e) A material used as a source of energy (e.g., Calor gas).
2. (a) Carbon and hydrogen.
   (b) Oxygen, nitrogen and sulphur. (Many other elements also occur in organic compounds.)
   (c) (i) Ether; (ii) Rubber.
   (d) 88 u $molecule^{-1}$.
3. One sequence taken from the chart on pages 32/33 is: crude oil → naphtha → ethene → plastics. Many other sequences are possible.
4. (a) Rubber and benzopyrene.
   (b) (i) Rubber and geraniol; (ii) These molecules contain C/C double bonds.
   (c)

–O–H	–C(=O)–O–H	–N(–H)–H
Geraniol	Benzoic acid	Putrescine

(d) Functional groups.

(e) Geraniol (two double bonds), or penicillin and aspartame (two carbonyl, >CO, groups), or putrescine and cadaverine (two amino groups).

5.–6.

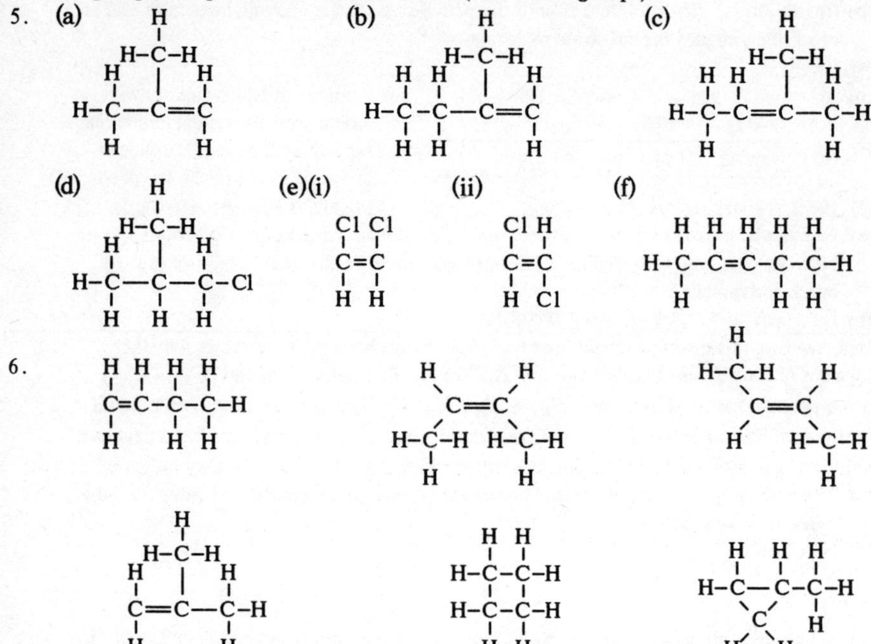

(Four alkenes: but-1-ene, cis-but-2-ene, trans-but-2-ene, 2-methylprop-1-ene; two cycloalkanes: cyclobutane, methylcyclopropane.)

7. (a) 2,4-dimethylpentane.
   (b) 2,3-dimethylpentane.
   (c) 2-methylbut-1-ene.
   (d) 2-methylbut-1-ene.
   (e) 2,2,4-trimethylpentane.
   (f) 2-methylbut-1-ene.

8.–9.

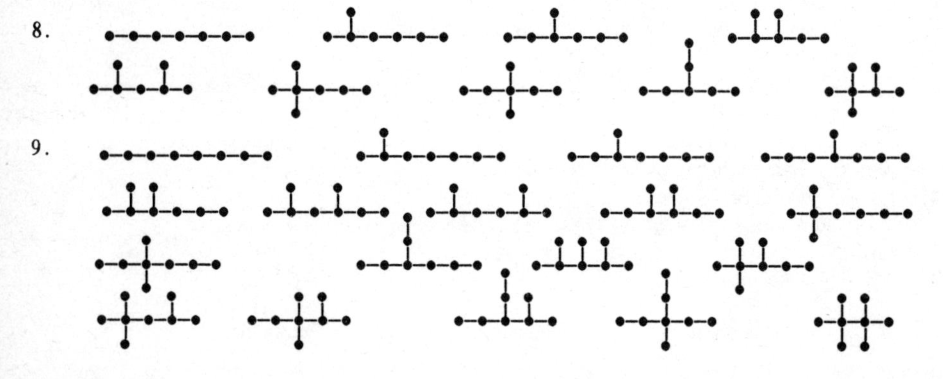

10. (a) Propan-1-ol. (b) Propan-2-ol.
    (c) Ethanoic acid. (d) Methylamine.
11. (a) CH_3COOH. (b)
    (c) 60.
    (d) Burns readily with a slightly luminous blue flame.
    (e) 7.
    (f) H_2 evolved.
    (g) No reaction.
    (h) Biomass (starches and sugars).
    (i) Air oxidation (of the naphtha fraction from crude oil).
    (j) Plastics (e.g., polyvinyl acetate, PVA). Other answers possible.
12. (a) (i) Head–tail–head.
    (ii) The two functional groups ($-NH_2$ and $-COOH$) per molecule are different.
    (iii) Polypeptides or protein.
    (iv) –NHCHXCO–NHCHXCO–NHCHXCO–
    (b) (i) The two functional groups of each monomer are the same.
    (ii) Bifunctional molecules in which the functional groups are the same are cheaper than those in which the groups are different.
    (iii) Polyamides.
    (iv) $-NH(CH_2)_6NH-CO(CH_2)_4CO-NH(CH_2)_6NH-$.
13. (a) Domestic and industrial fuel gas.
    (b) (i) About −164 °C.
    (ii) Liquefied natural gas.
    (c) (i) Natural gas liquids.
    (ii) Fractional distillation.
    (d) As a feedstock. (Cracking gives ethene for plastics manufacture.)
    (e) (i) Liquefied petroleum gas.
    (ii) Propane: cutting and welding metal. Butane: heating, lighting, cooking.
14. (a) Cracking is a reaction that breaks molecules into smaller molecules.
    (b)

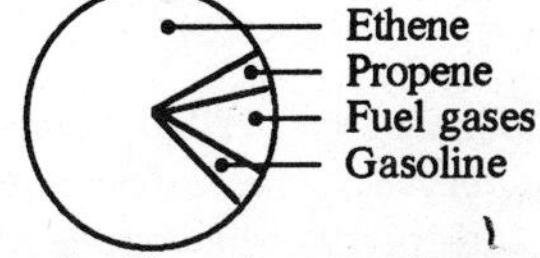

    (c) (i) $4C_2H_4 \rightarrow C_8H_{16}$.
    (ii) Addition (of hydrogen).
    (iii) $C_8H_{16} + H_2 \rightarrow C_8H_{18}$.
    (d) Propane is a larger molecule than ethane and can therefore give a greater variety of products.
    (e) $CH_3CH_2CH_3 \rightarrow CH_3CH{=}CH_2 + H_2$. (This reaction is also a dehydrogenation.)
15. (a) B.p. = −90 °C. (b) Liquid.
    (c) Propane molecules are smaller than butane molecules. Because of this the van der Waals' attraction between propane molecules is weaker. Liquid propane therefore boils at a lower temperature. (d) x bar is <31 bar; y bar is <6 bar.
16.

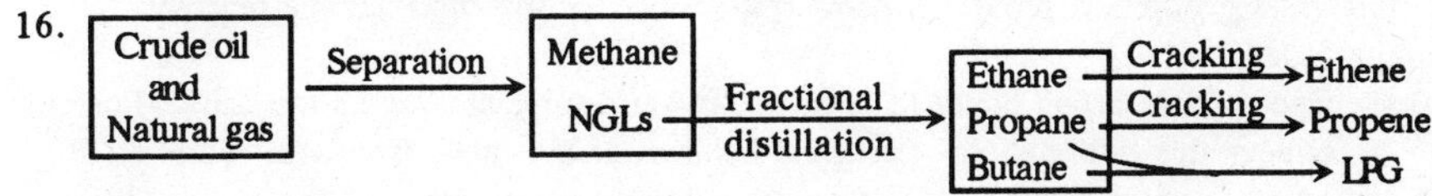

17. (i) Steam hinders the polymerisation of unsaturated molecules. (ii) Hydrogen reacts with unsaturated molecules to give saturated molecules. (iii) Catalysts allow the use of lower temperatures; catalysts determine the nature of the products.

18. (a) (i)

```
  H H H H H H H H
  | | | | | | | |
H-C-C-C-C-C-C-C-C-H
  | | | | | | | |
  H H H H H H H H
```

Octane

(ii) Four possible answers, one of which is

```
  H H H H H
  | | | | |
H-C-C-C-C-C-H
  | | | | |
  H | H H H
  H-C-H
    |
    H
```

2-methylpentane

(iii) There are nine isomers of C_6H_{12} (ignoring stereoisomers), one of which is

```
      H H
       \/
       C
    /     \
H>C         C<H
H           H
   |        |
H>C         C<H
H           H
    \     /
       C
       /\
      H H
```

Cyclohexane

(iv)

```
        H                            H
        |                            |
  H   C   H                   H    C    H
   \ / \ /                     \  / \\ /
    C   C                        C    C
    | O |          or           ||    |
    C   C                        C    C
   / \ / \                     /  \ // \
  H   C   H                   H    C    H
      |                            |
      H                            H
```

Benzene

(b) Straight-chain alkanes.

(c) (i) Changing the shape of a molecule. (The number of carbon atoms per molecule is usually unchanged but the hydrogen atoms are often decreased.)

(ii) Octane, a straight alkane, is reformed to a mixture that includes branched alkanes, cycloalkanes and aromatics. The following is one of many possible answers.

```
                                      H
                                      |
  H H H H H H H H              H  H-C-H H     H     H
  | | | | | | | |              |    |   |     |     |
H-C-C-C-C-C-C-C-C-H   -->    H-C----C---C-----C-----C-H
  | | | | | | | |              |    |   |     |     |
  H H H H H H H H              H  H-C-H H  H-C-H    H
                                    |         |
                                    H         H
```

(2,2,4-trimethylpentane)

(iii) Reforming creates hydrocarbons that burn better in spark-ignition car engines. (Branched/cyclic hydrocarbons have higher octane numbers. 2,2,4-trimethylpentane has an O.N. of 100 while octane has a negative O.N..)

(d) Heavy naphtha contains larger molecules than light naphtha. Intermolecular attraction is greater and the boiling point therefore higher. ('Inter' means 'between'.)

(e) Naphtha contains much larger molecules, which can give a much wider variety of products.

(f) (i) Hydration (or addition).

(ii) $C_2H_4 + H_2O \rightarrow C_2H_5OH$

19. A functional group is a group of atoms responsible for the characteristic reactivity of a compound.

20. (a) (i) The carbon/carbon bonds in benzene are very stable and are all the same. Formula A conveys this information. (Benzene behaves as if it were saturated. Formula B, therefore, is not preferred. However, formula B is not wrong, and can be useful in some circumstances. Imagine that the alternating single/double bond arrangement is mobile around the ring.)

(ii) Benzene reacts more readily by substitution than addition.

(b)

Ring A

or

Ring B

21. (a) Carbonisation, pyrolysis.
    (b) Coal gas, ammoniacal liquor and coke.
    (c) Toluene: C_7H_8; xylene: C_8H_{10}; pyridine: C_5H_5N; quinoline: C_9H_7N; phenol: C_6H_6O; cresol: C_7H_8O; naphthalene: $C_{10}H_8$; anthracene: $C_{14}H_{10}$.
    (d) Crude oil.
22. (a) Cyclohexene.
    (b) (i) Br Br (ii) 1,2-dibromocyclohexane. (iii) $C_6H_{10}Br_2$.
23. (a)

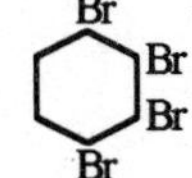

(b) 1,2,3,4-tetrabromocyclohexane. (c) $C_6H_8Br_4$.

24. (a) (i) Benzene. (ii) C_6H_6.
    (b) (i) Br or Br (ii) Bromobenzene (or phenyl* bromide) and hydrogen bromide. (iii) C_6H_5Br.
    *Phene is an obsolete name for benzene.
25. Q.22: addition; Q.23: addition; Q.24: substitution.
26. (a) F. (b) D. (c) C and E. (d) A. (e) B.
27. (a) Chloroethene, vinyl chloride.
    (b) Methylbenzene, toluene.
    (c) Benzenecarboxylic acid, benzoic acid.
28. (a) 1,2-dibromobutane. (b) 2,3-dibromobutane. (c) 1,2-dibromo-2-methylpropane.
29. (a) A: Fractional distillation (fractionation).
    B: Cracking.
    C: Dimerisation, the joining of two monomers to give a dimer. (Three monomers give a trimer. Four give a tetramer. More than four give a polymer.) The reaction is also an addition because one molecule adds across the double bond of the other.
    D: Addition polymerisation.

(b)

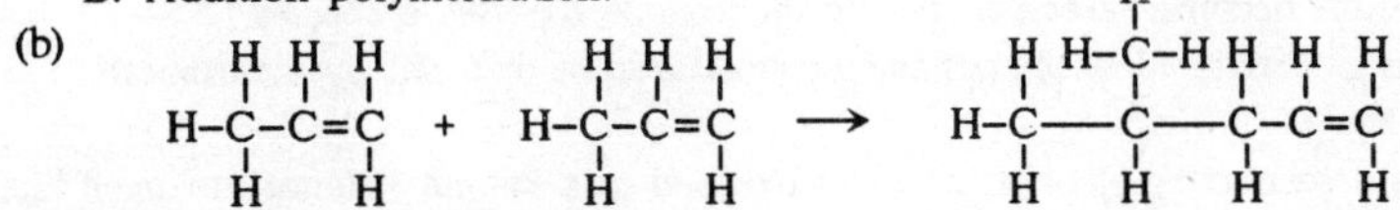

(c) Ethene and 4-methylpent-1-ene.

(d) (i) and (ii) (iii) The 2-methylpropyl group.

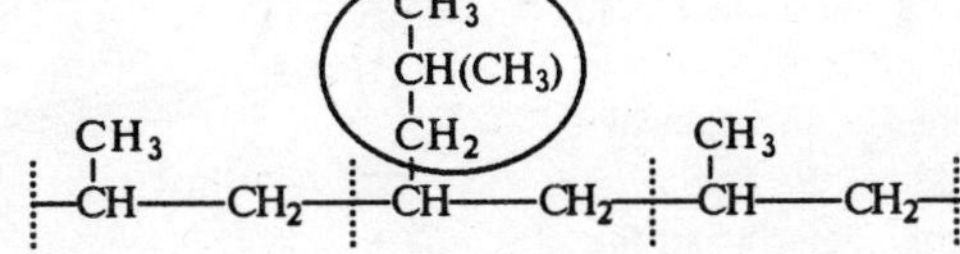

30. (a)

	Lignite	Bituminous coal	Anthracite
Age	Young	Middle-aged	Old
Carbon/%	60	80	90
Calorific value/kJ g^{-1}	20	32	34

(b) Sulphur. (c) Various answers, e.g., benzene.
('Volatile' means 'easily turns into a vapour; easily evaporates'.)

31. Ethanol: an alcohol (alkanol) with the formula C_2H_5OH.
Sugars: water-soluble carbohydrates with the formula $C_6H_{12}O_6$ or $C_{12}H_{22}O_{11}$.
Enzyme: an organic catalyst, which accelerates and directs reactions.

32. Butan-1-ol, $CH_3CH_2CH_2CH_2OH$ and butan-2-ol, $CH_3CH_2CHOHCH_3$.

33. (a) C. (b) B. (c) E. (d) A. (e) A. (f) A. (g) D. (h) E.

34. (a) (i) Dehydration; (ii) $C_2H_5OH \rightarrow C_2H_4 + H_2O$.
(b) Catalyst.
(c) Ethene.
(d) Bromine water test: ethene discharges the brown colour.
(e) (i) Propene; (ii)

$$\begin{array}{ccccccc} & H & & H & & H & \\ & | & & | & & | & \\ H- & C & - & C & - & C & -OH \\ & | & & | & & | & \\ & H & & H & & H & \end{array} \longrightarrow \begin{array}{ccccccc} & H & & H & & H \\ & | & & | & & | \\ H- & C & - & C & = & C \\ & | & & & & | \\ & H & & & & H \end{array} + H_2O$$

35. *Closed questions*
(a) B.
(b) E.
(c) (i) A; (ii) C.
(d) (i) D; (ii) D; (iii) B.
Open questions
(a) (i) A and D (2 marks); (ii) B and C (2 marks).
(b) C and F (2 marks).
(c) F (1 mark).

36. (a) 2-methylbutan-1-ol. (b) Butan-2-ol. (c) 2,2-dimethylpropan-1-ol.

37. (a) (i) B is Z;
(ii) A change of colour from orange to green. (Dichromate solution is orange. The solution becomes green when the dichromate oxidises a species.)
(iii) Z is a tertiary alcohol: tertiary alcohols are not oxidised by dichromate.
(b) (i) C is X.
(ii) X is a secondary alcohol, which is oxidised to a ketone. Hence the smell like acetone, which is also a ketone.
(c) (i) A is W; D is Y.
(ii) The data book gives the boiling point of butan-1-ol (W) as 118 °C.
(iii) The liquid is distilled, a thermometer being placed at the outlet of the flask. The reading on the thermometer when the liquid distils gives the boiling point.

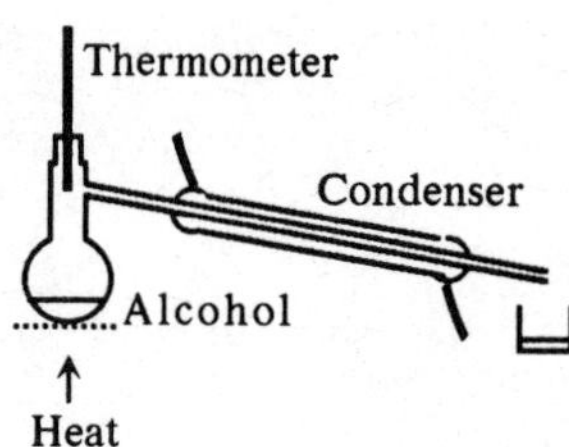

(iv) D (Y) is a more compact molecule than A (W); branched molecules are more compact than straight molecules. The van der Waals' attraction between compact molecules is weaker and the boiling point of the liquid therefore lower.

(d) W: butan-1-ol. X: butan-2-ol. Y: 2-methylpropan-1-ol. Z: 2-methylpropan-2-ol.

38. Glucose is an aldehyde (alkanal) because it contains the aldehyde group, –CHO. It therefore gives a positive result with Fehling's Reagent. Sucrose is not an aldehyde and gives a negative result with this reagent.

39. Methanoic acid is also an aldehyde: H–C(=O)–OH (the H–C=O group boxed)

40. (a) A: CH_3CH_2OH. B: CH_3CHO. C: CH_3COOH. D: $CH_3CHOHCH_3$.
E: CH_3COCH_3.

(b) (i) 2; (ii) This oxidation occurs with Fehling's Reagent while the other two oxidations do not. What this means is that aldehydes give a positive result with Fehling's Reagent, which oxidises the aldehyde to an acid. Alcohols give a negative result, as do acids and ketones.

(c) $CH_3CHOHCH_3 \xrightarrow[(-2H)]{\text{oxidation}} CH_3COCH_3$

41. (i) Butanone (butan-2-one is unnecessary); (ii) Methylbutanone (3-methylbutan-2-one is unnecessary).

42. $C_7H_{11}O$.

43. (a) A: coal. B: oil. C: natural gas.

(b) Steam reforming is a reaction in which a hydrocarbon is heated with steam to give a mixture of carbon monoxide and hydrogen. (Do not confuse steam reforming with reforming: reforming is a reaction that alters the shape of a hydrocarbon molecule. The number of carbon atoms per molecule remains the same though some hydrogen atoms may be lost.)

(c) Carbon monoxide and hydrogen. (Synthesis gas is often called syngas.)

44. (a) 1: $C(s) + H_2O(g) \rightarrow CO(g) + H_2(g)$; 3: $CH_4(g) + H_2O(g) \rightarrow CO(g) + 3H_2(g)$.

(b) From coke, $CO:H_2 = 1:1$; from methane, $CO:H_2 = 1:3$.

(c) Synthesis gas made from naphtha. (The $CO:H_2$ ratio is 1:2, which is the same as the volume ratio of reactants for making methanol.)

(d) A thermosetting plastic such as urea-methanal (urea-formaldehyde).

45. (a) Endothermic: heat taken in; exothermic: heat given out.

(b) The process can be run continuously because the reaction chamber maintains a steady temperature. (The heat absorbed by the endothermic reaction is supplied by the exothermic reaction.)

(c) Oxygen increases the proportion of carbon monoxide in the synthesis gas.

(d) $CO:H_2 = 3:1$. (Add eqns 1 and 2 to get:
$3C(s) + H_2O(g) + O_2(g) \rightarrow \underline{3}CO(g) + \underline{1}H_2(g)$.)

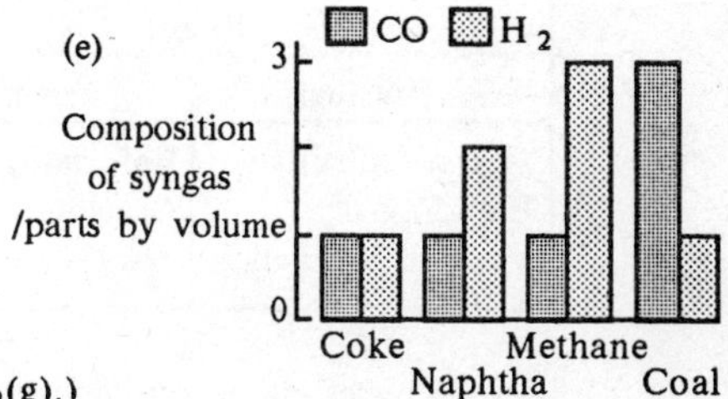

46. (a) (i) £189.75; (ii) £284.00.
    (b) Naphtha: £(189.75 − 65.00) = £124.75.
    Ethane: £(284.00 − 40.00) = £244.00.
    (c) From naphtha: £(124.75/0.3) = £415.83 profit per tonne.
    From ethane: £(244.00/0.8) = £305.00 profit per tonne.
    (d) From naphtha.
    (e) Other costs: labour (wages and salaries), transport, energy, insurance, provision for employee pensions, taxation, interest payments on loans, depreciation, replacement and renewal of plant, machinery and buildings. Select any four.
47. (a) Polystyrene. (b)

```
  [C6H5]     [C6H5]     [C6H5]
    |          |          |
 —CH—CH2—CH—CH2—CH—
```

48. Other factors: effect on the environment, safety, employee welfare, government policies. Select two.
49. (a) (i) Ethanol (ethyl alcohol); (ii) ethanoic acid (acetic acid).
    (b) Allow air to enter.
    (c) They have a large surface area.
    (d)

```
     H H                        H
     | |                        |   //O
   H-C-C-OH  +  O2   -->      H-C-C         +  H2O
     | |                        |   \OH
     H H                        H
```

50.

```
   H H
   | |  //O
 H-C-C-C
   | |  \OH     propanoic acid
   H H
```

51.

```
      //O            //O
  H-C               C
      \OH           |  \OH
                    |  //O
                    C
                       \OH
```

Methanoic acid, formic acid.

Ethandioic acid, oxalic acid.

Monoprotic acids have one replaceable hydrogen atom per molecule, diprotic have two, triprotic, three, and so on.

52. (i) 2-methylbutanoic acid; (ii) 4-methylpentanoic acid.
53. (a) ***Processes used to make petrol and diesel***

Petrol		Diesel	
Physical	Chemical	Physical	Chemical
Primary distillation Blending Vacuum distillation	Reforming Cracking Alkylation	Primary distillation Blending	None

    (b) More processes are needed to make petrol than to make diesel. Furthermore, some of these are chemical, and chemical processes are costlier than physical processes.

54. (a) Catalyst.

(b) It is recycled.

(c) (i) The given molecules are $2C_8H_{16}$, C_7H_{16} and C_4H_8. These account for 27C and 56H. The other molecule, therefore, is C_3H_6, which is propene. (It is unlikely to be cyclopropane.)

(ii)

$$\begin{array}{ccccccc} & & H & & H & & \\ & & | & & | & & \\ H & - & C & = & C & - & C & - & H \\ & & & & | & & | \\ & & & & H & & H \end{array}$$

(iii) Polypropene (an addition polymer used in carpets, fishing nets, etc.).

(d) 2,2,4-trimethylpentane.

(e)

$$\begin{array}{ccccccc} & & CH_3 & & CH_3 & & \\ & & | & & | & & \\ CH_3 & - & C & — & C & - & CH_3 \\ & & | & & | & & \\ & & CH_3 & & CH_3 & & \end{array}$$

(f) Cracked spirit consists of straight-chain hydrocarbons. These do not burn smoothly in spark-ignition (petrol) engines; they burn too quickly (detonate) causing the engine to 'pink', 'knock', 'rattle'. This wastes energy and may damage the engine. Alkylate consists of branched-chain hydrocarbons. These burn smoothly.

55. (a)

Petrol	Diesel
Gasoline fraction	Gas oil fraction
Reformed naphtha	Kerosine†
Cracked spirit	
Alkylate	
Tetraethyl lead *	
Butane †	

*Lead-free petrol is becoming more common. When tetraethyl lead is added, dibromoethane, CH_2BrCH_2Br, is also added. This removes lead from the engine.

†But only in winter.

(b) Petrol engines mix petrol *vapour* with air. This is possible because petrol evaporates quickly: it is volatile. But not in cold weather. So butane is added because it evaporates quickly in cold weather.

(c) (i) Kerosine; (ii) Diesel oil freezes in very cold weather. Kerosine has a lower freezing point than diesel oil and remains a liquid.

56. (a) In petrol engines, a mixture of petrol vapour and air is compressed and ignited by a spark. In diesel engines, air is compressed (twice as much as in petrol engines), which makes it very hot. Liquid diesel oil is then injected (as a fine spray) and is ignited by the hot compressed air.

(b) (i) Petrol: 8; diesel oil: 16.

(c) Petrol engines need branched and cyclic hydrocarbons (to burn slowly after sparking). Diesel engines need straight-chain hydrocarbons (to burn instantly over the period of injection).

57. (a) Methane. (Marsh gas and firedamp are also methane.)

(b) Digestion = fermentation by anaerobic bacteria.

(c) First: methane/air mixtures are explosive. Second: air (oxygen) kills the bacteria.

(d) It is used to heat (30 °C) the digesters.

(e)

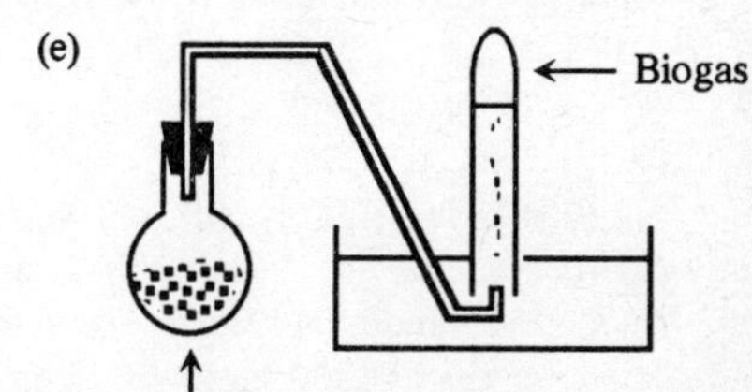

Unit 2 Answers

58. (a) Add amine to copper sulphate solution. A pale blue precipitate is obtained. This redissolves on further addition of amine to give a deep blue solution.

(b) Methylamine

```
   H   H
   |  /
H–C–N
   |  \
   H   H
```

(c) $C_2H_5NH_2 + H_2O \rightarrow C_2H_5NH_3^+ + OH^-$
ethylammonium ion

59. (a) The plastic cannot be resoftened after it has set.
(b) Because the bonds formed as it sets are in three dimensions (left/right, up/down, back/front). This creates a giant macromolecule/network solid.
(c) (i) The drawing gives the false impression that polymerisation occurs in only two directions (dimensions).
(ii) By drawing some of the bonds above and below the paper using the convention shown opposite.
(d) Various answers. Good insulator. Electrical switches.
(e)

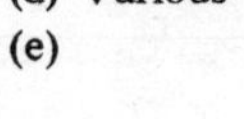

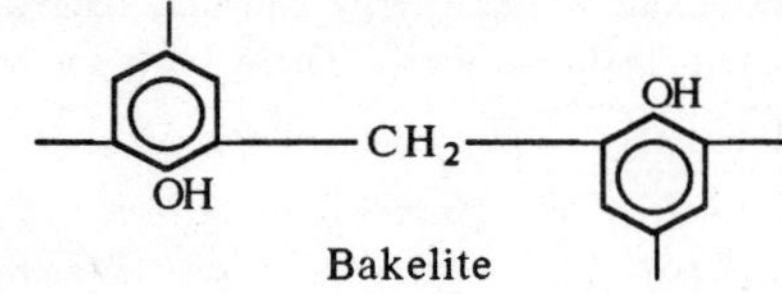

Bakelite

60. (i) Aldehyde (alkanal); (ii) Alkane; (iii) Ketone (alkanone); (iv) Acid (carboxylic acid, alkanoic acid); (v) Alcohol (alkanol); (vi) Alkene; (vii) cycloalkane; (viii) Acid and aldehyde.

Unit 3

1. (a) Kilogram. (b) *V*. (c) m^3. (d) Space occupied. (e) mol. (f) Charge. (g) Coulomb. (h) Charge is a property of matter that may be positive or negative.
2. (a) (i) 12 g; (ii) 64 g; (iii) 197 g; (iv) 1 g; (v) 2 g; (vi) 4 g; (vii) 16 g; (viii) 32 g; (ix) 48 g; (x) 160 g.
(b) (i) 18 g; (ii) 98 g; (iii) 58.5 g; (iv) 46 g.
(c) (i) 23 g; (ii) 35.5 g; (iii) 18 g; (iv) 96 g.
(d) The mass of an electron is about $^1/_{2000}$ u. A mol of e^- therefore has a mass of about $^1/_{2000}$ g.
3. Seven elements: hydrogen, nitrogen, oxygen, fluorine, chlorine, bromine and iodine. (Eight if we include astatine, but this element is almost non-existent.)
4. (i) 342 g mol^{-1}; (ii) 66 560 g mol^{-1}.
5. Relative molecular mass is 34, hence formula must be PH_3.
6. 10^{78}. 7. 1.9 x 10^{16} yr. 8. (i) 44 u; (ii) 44; (iii) 44 g; (iv) 44 g mol^{-1}.
9. (a) (i) 24 g; (ii) 160 g; (iii) 12.4 g; (iv) 8.1 x 10^{-4} g.
(b) (i) 32 g; (ii) 190 g.
(c) (i) 400 mg (0.4 g); (ii) 800 kg (800 000 g).
10. (a) (i) 0.5 mol; (ii) 1 mol; (iii) 2 mol; (iv) 0.75 mol.
(b) (i) 0.25 mol; (ii) 2.08 x 10^{-5} mol.
(c) (i) 1 kmol (1000 mol); (ii) 1 mmol (0.001 mol).

Unit 3 Answers

11. (a) To carry the products of decomposition through the U-tubes.
    (b) (i) Anhydrous calcium chloride; (ii) calcium oxide or soda-lime.
    (c) (i) 0.015 mol; (ii) 0.03 mol; (iii) 2:1; (iv) 2:1; (v) $(PbCO_3)_2.Pb(OH)_2$ or $Pb_3(CO_3)_2(OH)_2$.
12. $w = 10$, $x = 5$, $y = 9$, $z = 1/2$.
13. (a) 2 mol. (b) 0.4 mol L^{-1}. (c) 4 L. (d) 0.1 mol L^{-1}. (e) 3 mol. (f) 6 L.
14. (a) 1.67 mol L^{-1}. (b) 1.25 mol L^{-1}. (c) 1.75 mol L^{-1}. (d) 5.5 mol L^{-1}.
15. (a) 0.5 mol L^{-1}. (b) 0.25 mol L^{-1}. (c) 0.09 mol L^{-1}. (d) 0.125 mol L^{-1}.
    (e) 0.17 mol L^{-1}.
16. (i) Atoms; (ii) atoms; (iii) ion-pairs; (iv) molecules; (v) ions; (vi) electrons.
17. (a) 1.2×10^{23} atoms. (b) 1.2×10^{23} ions. (c) 1.17×10^{9} molecules.
    (d) (i) 1.56×10^{26} e^-; (ii) 1.44×10^{25} e^-. (e) 6×10^{23} e^-. (f) 101 g. (g) 29.4 g.
    (h) (i) 0.25 mol; (ii) 1.5×10^{23} molecules; (iii) 4.5×10^{23} atoms. (i) 1/2 mol.
    (j) 5×10^{22} atoms. (k) 3.33×10^{-24} g. (l) 1.82×10^{13} J. (m) 7.25×10^{-19} J.
18. 7.9×10^{10} molecules L^{-1}.
19. (a) $4.49 \times 10^{-2} \times (10^{-7})^3$ cm^3 = 4.49×10^{-23} cm^3 .
    (b) Volume = Mass/Density, so volume of 1 mol = 58.5 g/2.17 g cm^{-3} = 26.96 cm^3.
    (c) N_A = (26.96 cm^3)/(4.49×10^{-23} cm^3) = 6.0×10^{23}.
20. Density = 8.92 g cm^{-3}. R.a.m. (A_r) = 64. Hence V of 1 mol = 7.17 cm^3, which consists of 6×10^{23} atoms. V of block = 60 cm^3, which consists of 5×10^{24} atoms.
21. (a) 1/2 mol. (b) 16 g. (c) 8g (not 4 g)!
22. (a) 1:2. (b) Nitrogen, N_2, or ethene, C_2H_4. (c) $15N_A$. (d) 6×10^{24} molecules.
    (e) 14 g. (f) 1.5×10^{23} Na^+. (g) 6×10^{17} atoms.
23. 0.0068 cm^3 contains 1.7×10^{17} atoms ... 24 000 cm^3 contains 6×10^{23} atoms.
    Hence L = 6×10^{23} (atoms) mol^{-1} (of helium).
24. (a) $V_m(CH_4)$ = 24.82 L mol^{-1},
    $V_m(N_2)$ = 24.80 L mol^{-1},
    $V_m(C_3H_8)$ = 24.79 L mol^{-1}.
    (b) Greater.
    (c) R.t.s.p. uses a higher temperature and lower pressure than s.t.p., both of which changes increase the volume of a gas.
    (d) From graph, V_m at 20 °C is 24.3 L mol^{-1}.
    (e) That there is a straight line relationship between V_m and T.

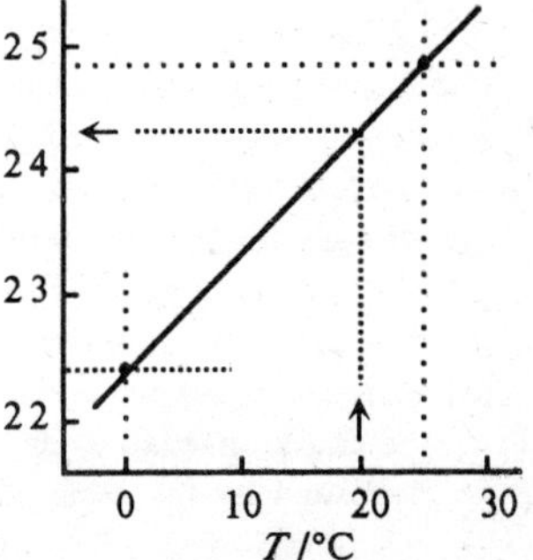

25. 23.24 g occupies 10.30 cm^3. 12 g occupies 5.32 cm^3. V_m = 5.32 cm^3 mol^{-1}.
26. (a) 1.13 g L^{-1}. (b) 24.7 dm^3 mol^{-1}. (c) 2.04 g L^{-1}.
27. (a) Molecular size increases by a CH_2 group through the series: CH_3OH ... CH_3CH_2OH ... $CH_3CH_2CH_2OH$ ... $CH_3CH_2CH_2CH_2OH$.
    (b) The increase in V_m from one member to the next is V_m of 'CH_2'. Hence $V_m(CH_2)$ is 18.0, 16.4 and 16.7 mL mol^{-1}, giving an average V_m of 17.0 mL mol^{-1}.
    (c) The series continues after butanol with pentanol and hexanol. V_m(hexanol) will therefore be approximately [91.5 + (2 x 17.0)] mL mol^{-1} = 125.5 mL mol^{-1}.

28. (a) 3/4. (b) 7/11. (c) 5/4. (d) 2/5.
29. 0.0596 g N_2 occupies 50 cm^3. $M_m(N_2)$ = 28 g mol^{-1}. 28 g N_2 occupies 23490 cm^3, which is the volume of 1 mol.
30. (a) Molar volumes increase because atom size increases from Li to Na to K.
    (b) Molar volumes increase because ion size increases from Cl^- to Br^- to I^-.
    (c) Carbon atoms in diamond are closer and so diamond's molar volume is smaller.
    (d) Because sugar molecules are much larger than any of the above atoms or ions.
    (e) $M_m = V_m \times$ Density. Density of graphite = (12 g mol^{-1})/(5.3 cm^3 mol^{-1}) = 2.26 g cm^{-3}; density of diamond = 3.53 g cm^{-3}. The greater density of diamond confirms that the atoms in diamond are more closely packed than in graphite.
31. (a) 12.4 L. (b) 11 g.
32. $H_2O_2(aq) \rightarrow H_2O(l) + \frac{1}{2}O_2(g)$. 1 mol H_2O_2 gives ½ mol O_2, for which V = 12 L.
33. (a) (i) 2 g; (ii) 4 g; (iii) 28 g.
    (b) (i) 26 g lighter; (ii) 26 g.
    (c) (i) 24 g lighter; (ii) 24 g.
    (d) No. The hydrogen-filled balloon can lift only 26 g compared with the helium-filled balloon's 24 g, i.e., only 2 g more.
    (e) (i) Worked out from knowledge with pen and paper.
    (ii) Practical/experimental. Set up the balloons and measure the loads they can lift.
34. (a) 400 C. (b) 6 A. (c) 11 s.
35. $V = 6 \times 10^{23} \times (0.014 + 0.016) \times (10^{-7})^3$ cm^3 = 18 cm^3.
36. $Q = (10^{-10}$ A x 1 s) = 10^{-10} C. Dividing by 1.6×10^{-19} C gives 6.25×10^8 e^-.
37. $Sn^{2+} + 2e^- \rightarrow Sn$

    2 mol → 1 mol

    Q = 193 000 C, m = 119 g

    1.19 g Sn needs 1930 C in 1200 s. Hence, I = 1.61 A.
38. For a given substance, mass is proportional to amount.
39. A singly-charged ion needs 96 500 C per mole of metal deposited but a doubly charged ion needs 2 x 96 500 C, a triply charged ion, 3 x 96 500 C: $M^{z+} + ze^- \rightarrow M$ where z is the 'charge number' of the ion.
40. (a) Cell A: $Ag^+(aq) + e^- \rightarrow Ag(s)$. Cell B: $Cu^{2+}(aq) + 2e^- \rightarrow Cu(s)$. ($Cu^{2+}(aq)$ is discharged more easily than $H^+(aq)$.)
    (b) (i) Mass of copper = (0.032/108) x 64 x 1/2 g = 0.0095 g. Alternatively, work out the charge needed to deposit 0.032 g Ag and then work out the mass of copper deposited by the same charge.
    (ii) (0.0091/0.0095) x 100% = 95.79%.
    (c) For 0.032 g Ag, Q = 28.6 C. Scaled up to 108 g Ag, which required an Avogadro number of electrons, Q = 96 525 C. Dividing by 1.6×10^{-19} C gives N_A as 6.03×10^{23}. Hence, $L = 6.03 \times 10^{23}$ mol^{-1}.
41. (a) (i) $2H^+(aq) + 2e^- \rightarrow H_2(g)$

    2 electrons → 1 molecule

    N_A electrons → $\frac{1}{2}N_A$ molecules

    n = ½ mol

    V = 12.4 L

    (ii) $2H^+(aq) + 2e^- \rightarrow H_2(g)$

    2 mol → 1 mol

    Q = 193 000 C, V = 24.8 L

    Q used = 54 C

    $V(H_2)$ = (24.8/193 000) x 54 L = 6.9×10^{-3} L (4.6 mL)

(b) $2OH^-(aq) \rightarrow H_2O(l) + \frac{1}{2}O_2(g) + 2e^-$

$\frac{1}{2}$ mol 2 mol

V = 12.4 L Q = 193 000 C

(i) 6.2 L; (ii) Q used = 30 C, giving $V(O_2)$ = (12.4/193 000) x 30 L = 1.9×10^{-3} L.

42. (a) Cathode: $Na^+ + e^- \rightarrow Na$. Anode: $Cl^- \rightarrow \frac{1}{2}Cl_2 + e^-$.

(b) (i) Q = 9000 C; (ii) (23/96 500) x 9000 g of Na = 2.15 g of Na;

(iii) (12.4/96 500) x 9000 dm^3 = 1.156 dm^3 (1.156 L, 1156 cm^3, 1156 mL).

43. (a) (i) $Cu^{2+}(aq) + 2e^- \rightarrow Cu(s)$; (ii) $Cu(s) \rightarrow Cu^{2+}(aq) + 2e^-$.

(b) The same current passes through all cells in series. For one cell, m(Cu) deposited = 98 kg. Now 64 g Cu needs 193 000 C. So 98 kg needs 2.96×10^8 C = Q. Time (t) = 14 d = 1.21×10^6 s. From $Q = It$, current = 244.6 A.

(c) (i) Silver and gold; (ii) They are in the same group of the Periodic Table and so have similar chemistry.

(d) (2.1/2.3) x 100% = 91.3%.

44. Q used = 2395.2 C, which deposits 0.97 g, i.e., 96 500 C deposits 39 g. If the metal ion is M^+ (i.e., singly charged), then r.a.m. is 39 and the metal is K. (If the ion were M^{2+}, r.a.m. would be 19.5; if M^{3+}, 13; if M^{4+}, 9.75. No metals in the Periodic Table have these values.)

45. (a) 6 x 96 500 C = 579 000 C. (b) 6 mol.

46. Q used = 175 000 A x 86 400 s = 1.512×10^{10} C, passing through 80 cells in series.

$Al^{3+} + 3e^- \rightarrow Al$

3 mol 1 mol

Q = 289 000 C m = 27 g ... from which 1.512×10^{10} C give 1412.6 kg d^{-1} $cell^{-1}$.

Total mass of aluminium = 1412.6 x 80 kg d^{-1} = 113 008 kg d^{-1} = 113.01 t d^{-1}.

47. (a) (i) $H_2(g) + \frac{1}{2}O_2(g) \rightarrow H_2O(l)$

(ii) 1 mol $\frac{1}{2}$ mol 1 mol

1 vol. $\frac{1}{2}$ vol. Negligible

10 mL + 5 mL → -

Product composition: a trace of water plus 13 mL of unused (excess) oxygen.

(b) (i) $CO(g) + \frac{1}{2}O_2(g) \rightarrow CO_2(g)$

(ii) 1 mol $\frac{1}{2}$ mol 1 mol

1 vol. $\frac{1}{2}$ vol. 1 vol.

20 mL + 10 mL → 20 mL

Product composition: 20 mL carbon dioxide plus 27 mL excess oxygen.

(i) Equation	(ii) Product composition
(c) $CH_4(g) + 2O_2(g) \rightarrow CO_2(g) + 2H_2O(l)$	30 mL CO_2, 14 mL O_2, trace $H_2O(l)$.
(d) $C_2H_4(g) + 3O_2(g) \rightarrow 2CO_2(g) + 2H_2O(l)$	80 ml CO_2, 90 mL O_2, trace $H_2O(l)$.
(e) $C_3H_4(g) + 4O_2(g) \rightarrow 3CO_2(g) + 2H_2O(l)$	150 mL CO_2, 130 mL O_2, trace of $H_2O(l)$
(f) $2NH_3(g) + 1\frac{1}{2}O_2(g) \rightarrow N_2(g) + 3H_2O(l)$	30 mL N_2, 100 mL O_2, trace of $H_2O(l)$

(iii) The product compositions are the same in respect of CO_2 and excess O_2. But H_2O must now be treated as a gas (steam). The steam has the following volumes: (a) 10 mL, (b) Nil, (c) 60 mL, (d) 80 mL, (e) 100 mL, (f) 90 mL.

48. (a) $C_4H_{10}(g) + 6\frac{1}{2}O_2(g) \rightarrow 4CO_2(g) + 5H_2O(?)$.

(b) 10 mL 65 mL 40 mL Negligible volume of liquid at T = 99 °C.

50 mL of steam at T = 101 °C.

Gas composition. At 99 °C: 40 mL CO_2, 25 mL excess O_2.

At 101 °C: 40 mL CO_2, 25 mL excess O_2, 50 mL H_2O as steam.

49. (a) (i) $Mg + 2HCl \rightarrow MgCl_2 + H_2$; (ii) $Mg(s) + 2H^+(aq) \rightarrow Mg^{2+}(aq) + H_2(g)$.
(b) 1 mol 2 mol 1 mol
24 g 1000 mL (soln.) 24.8 L (gas)
2.4 g 100 mL 2.48 L
100 mL 2 mol L^{-1} acid reacts with 2.4 g Mg. But there are 5 g Mg, so Mg is in excess.
(c) 2.4 g Mg, reacting with 100 mL 2 mol L^{-1} HCl, give 2.48 L H_2.

50. $H_2O_2 \rightarrow H_2O + \frac{1}{2}O_2$. 1 L of 1 mol L^{-1} solution contains 1 mol H_2O_2. This gives $\frac{1}{2}$ mol O_2, which has a volume of 11 540 mL. V_m, therefore, is 23 080 mL mol^{-1}.

51. (i) $C_2H_6 + 2\frac{1}{2}O_2 \rightarrow 2CO + 3H_2O$. 2 L ethane require 5 L oxygen.
(ii) $C_2H_6 + 3\frac{1}{2}O_2 \rightarrow 2CO_2 + 3H_2O$. 2 L ethane require 7 L oxygen.

52. $C_xH_y(g) + ?O_2(g) \rightarrow xCO_2(g) + {}^y/_2H_2O(g)$.
10 mL 40 mL 40 mL
1 mol 4 mol 4 mol Therefore x = 4, y = 8. Formula = C_4H_8.

53. (a) 130 mL. (b) 30 mL.

54. $C_xH_y(g) + zO_2(g) \rightarrow xCO_2(g) + {}^y/_2H_2O(l)$.
9 mL 45 mL 27 mL –
1 vol. 5 vol. 3 vol. –
1 mol 5 mol 3 mol – ... which gives x = 3, z = 5.
The equation now is:
$C_3H_y + 5O_2 \rightarrow 3CO_2 + {}^y/_2 H_2O$.
In order to balance the equation in oxygen we must have $4H_2O$ on the right hand side. The value of $^y/_2$ is therefore 4 and so y is 8. Hence formula is C_3H_8.

55. (a) $B_2H_6(g) + 3O_2(g) \rightarrow B_2O_3(s) + 3H_2O(l)$.
(b) Small quantities of solid boron oxide and liquid water and 40 mL excess oxygen.

56. (a) Reaction 1: 1:5. Reaction 2: 1:5. Reaction 3: 1:1.
(b) 92 g C_7H_8 should give 78 g C_6H_6. Therefore 1 t should give 0.85 t.
% yield = (0.77/0.85) x 100% = 90.6 %.

57. (a) $BiO^+ + 2H^+ + 3e^- \rightarrow Bi + H_2O$. (b) $2HOBr + 2H^+ + 2e^- \rightarrow Br_2 + 2H_2O$.
(c) $BrO_3^- + 6H^+ + 5e^- \rightarrow \frac{1}{2}Br_2 + 3H_2O$. (d) $FeO_4^{2-} + 8H^+ + 3e^- \rightarrow Fe^{3+} + 4H_2O$.
(e) $NO + H_2O \rightarrow HNO_2 + H^+ + e^-$. (f) $PH_3 \rightarrow P + 3H^+ + 3e^-$.
(g) $H_3PO_3 + H_2O \rightarrow H_3PO_4 + 2H^+ + 2e^-$. (h) $Pb^{2+} + 2H_2O \rightarrow PbO_2 + 4H^+ + 2e^-$.
(i) $SO_4^{2-} + 2H^+ + 2e^- \rightarrow SO_3^{2-} + H_2O$. (j) $2S_2O_3^{2-} \rightarrow S_4O_6^{2-} + 2e^-$.
Reductions: (a), (b), (c), (d), (i); oxidations: (e), (f), (g), (h), (j).
(H^+ and H_2O have been used to construct the above equations. But OH^- may also be used. (e), for example, could be $NO + OH^- \rightarrow HNO_2 + e^-$.)

58. (a) In general: $E + H_2O \rightarrow EO + 2H^+ + 2e^-$ where E = Be, Mg, Ca, Sr, Ba.
(b) In general: $E + 2H^+ + 2e^- \rightarrow H_2E$ where E = S, Se, Te, Po. If E is oxygen, the reduction is $\frac{1}{2}O_2 + 2H^+ + 2e^- \rightarrow H_2O$.
(c) $H_2O_2 \rightarrow O_2 + 2H^+ + 2e^-$. (d) $H_2O_2 + 4H^+ + 4e^- \rightarrow H_2 + 2H_2O$.
(e) $H_2O_2 + 2H^+ + 2e^- \rightarrow 2H_2O$.

59. (a) $Zn + 2H^+ \rightarrow Zn^{2+} + H_2$. (b) $Zn + Cu^{2+} \rightarrow Zn^{2+} + Cu$.
(c) $Zn + 2Fe^{3+} \rightarrow Zn^{2+} + 2Fe^{2+}$. (d) $Zn + Br_2 \rightarrow Zn^{2+} + 2Br^-$.
(e) $H_2 + Cl_2 \rightarrow 2H^+ + 2Cl^-$ (f) $2Fe^{2+} + Cl_2 \rightarrow 2Fe^{3+} + 2Cl^-$.
(g) $MnO_4^- + 8H^+ + 5Fe^{2+} \rightarrow Mn^{2+} + 4H_2O + 5Fe^{3+}$.

60. The relation $V_1c_1/n_1 = V_2c_2/n_2$ may be used to solve titrimetric problems in general.
    (a) $c = c_2 = 0.042$ mol L^{-1}.
    (b) $c = c_2 = 0.112$ mol L^{-1}.
    (c) $c = c_1 = 0.098$ mol L^{-1}.
    (d) $c = c_1 = 0.033$ mol L^{-1}.
61. (a) A mass near to 6 g was weighed off (anything between 5.8 and 6.2 g will do). But it was weighed off so that the exact mass was known.
    (b) A weighing bottle containing the iron compound is weighed accurately (x g). The compound is tipped into a beaker. The bottle is reweighed accurately (y g). The mass of compound in the beaker is the difference between the weighings, $(x - y)$ g.
    (c) To remove dissolved oxygen, which would oxidise iron(II) to iron(III).
    (d) Because MnO_4^-(aq) ion needs H^+(aq) when it oxidises. (See ion-electron equation.)
    (e) MnO_4^-(aq), which is purple, oxidises Fe^{2+}(aq) and is itself reduced to Mn^{2+}(aq), which is (almost) colourless. At the end-point, the oxidation is complete. The first drop of excess MnO_4^-(aq) imparts a pink colour to the contents of the flask and the titration is stopped. No indicator needs to be added.
    (f) 17.45 mL.
    (g) (i) $V_1c_1/n_1 = V_2c_2/n_2$. $V_1 = 17.45$ mL; $c_1 = 0.05$ mL L^{-1}; $n_1 = 1$ mol; $V_2 = 20$ mL; $c_2 = ?$; $n_2 = 5$ mol. From these data, $c_2 = 0.218$ mol L^{-1};
    (ii) 60.69 g; (iii) 278.4 g; (iv) $(152 + 18x)$ g; (v) $152 + 18x = 278.4$, from which $x = 7.02$; (vi) $FeSO_4.7H_2O$.

Unit 4

1. (a) The substance formed when fatty acids react with glycerol in a mole ratio of 3:1.
   (b) (i) A; (ii) C; (iii) C.
   (c) (i) B;
   (ii) The molecule contains two double bonds;
   (iii) Addition/hydrogenation;
   (iv) $CH_2OCOC_{17}H_{35}$
   |
   $CHOCOC_{15}H_{31}$
   |
   $CH_2OCOC_{15}H_{31}$
   (d) (i) Saturated fat; (ii) Heart disease (narrowing of arteries); (iii) By replacing saturated fat in the diet with unsaturated/polyunsaturated fats and oils.
   (e) (i) Secondary; (ii) $C_{27}H_{46}O$.
2. (a)

Origin		
Plant	Animal	
	Land	Marine
Palm kernel oil	Lard	Cod liver oil
Olive oil	Tallow	Herring oil
Rapeseed oil	Suet	Halibut oil
Castor oil	Butter	
Linseed oil		
Corn oil		

   (b) Plants: liquids (oils); land animals: solids (fats); marine animals: liquids (oils).
3. (a) Various answers, e.g., hexane, petroleum ether, trichloroethane. (Any solvent that is water-immiscible, volatile and preferably non-poisonous.)
   (b) When the solvent is flammable.

(c) Because the seeds are exposed to fresh solvent after solvent and extract siphon over; and the cycle can be repeated as often as required without increasing the total volume of solvent.

(d) Take a mL or so of the solvent, place it on a filter paper and allow to evaporate. A translucent stain (of fat/oil) on the paper when viewed against the light means that extraction is not complete.

(e) Transfer the contents of the flask with rinsings to a weighed evaporating dish. Place the dish on an electric steam bath (in a fume cupboard) and evaporate to constant mass. The increase in mass over that of the empty dish is the mass of the fat/oil extract left in the dish.

4. (a) Precursor: a compound that gives rise to other compounds.
Essential fatty acids: acids that must be present in food because the body cannot make them from other substances in the food.
Polyunsaturated acids: fatty acids with two or more carbon/carbon double bonds per molecule. (Some fish oil fatty acids contain as many as five.)

(b) Octadec-9,12-dienoic acid ... 18 Cs in total ...two double bonds, one at C9 and one at C12 (numbering from the carboxylic C as 1). Hence linoleic acid is:

$$\overset{12}{CH_3(CH_2)_4CH=CHCH_2CH}=\overset{9}{CH}(CH_2)_7\overset{1}{C}OOH.$$

Octadec-9,12,15-trienoic acid. Similar to the above acid but with a third double bond at C15. Hence linolenic acid is:

$$CH_3CH_2CH= CHCH_2CH=CHCH_2CH=CH(CH_2)_7COOH.$$

5. (a) The number of double bonds per molecule.

(b) (i) Spike graph. See also Q. 10 of Unit 5.

(ii)

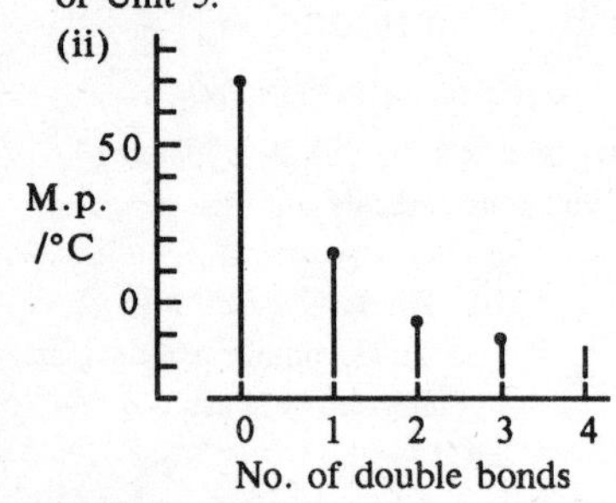

(iii) −13 °C.

(c) (i) Stearic acid molecules are straight and are able to pack closely. Van der Waals attraction then holds the molecules together and the acid has a high m.p.. Oleic acid molecules have a bend at the double bond. They cannot pack as closely. Van der Waals attraction is weaker, the acid has a lower m.p..

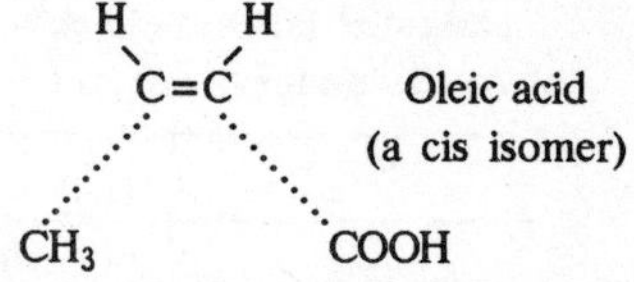

(ii) Oleic acid has a cis configuration at the double bond, which causes the bend in the molecule. Elaidic acid, in contrast, has a trans configuration and is a straight molecule despite the double bond. Packing is closer than in oleic acid and m.p. higher.

COOH
H
C=C
H
CH_3
Elaidic acid
(a trans isomer)

6. (a) $CH_2OCOC_{17}H_{35}$
$CHOCOC_{17}H_{35}$
$CH_2OCOC_{17}H_{35}$
Tristearin (glyceryl tristearate)

$CH_2OCOC_{17}H_{33}$
$CHOCOC_{17}H_{33}$
$CH_2OCOC_{17}H_{33}$
Triolein (glyceryl trioleate)

(b) Dissolve tristearin in hexane in a test tube. Repeat for triolein. Add a little bromine water to each solution, stopper and shake. Nothing happens with the tristearin because it is saturated. But the brown colour of bromine is rapidly discharged by the triolein because it is unsaturated.

(c) Tristearin would have a higher m.p. than triolein.

7. (a) Monoglyceride: only one of the three hydroxyl groups per molecule of glycerol has condensed (esterified) with a fatty acid.
Fully hardened fat = fully saturated fat: all double bonds have been removed by addition of hydrogen (hydrogenation).

(b)

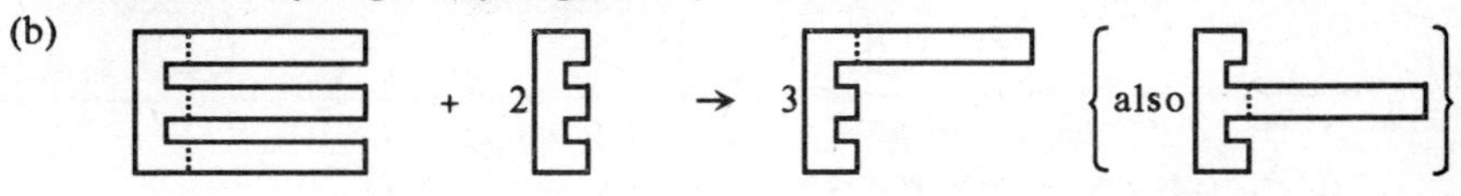

(c)

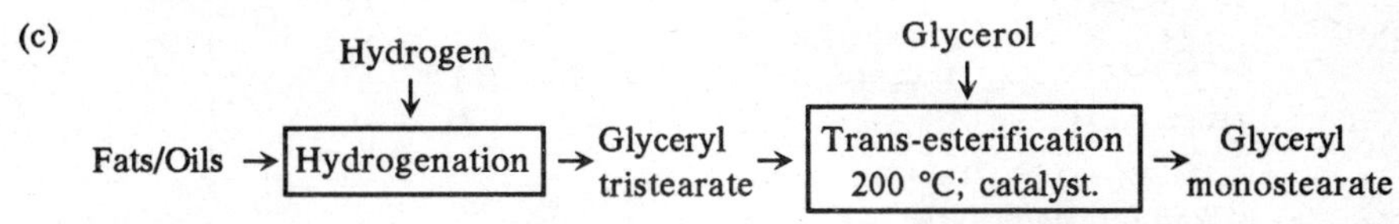

8. (a) (i) Because plant oils were plentiful while the supply of animal fats was limited.
(ii) Nickel. (iii) Complete hydrogenation would raise the m.p. well above body temperature (37 °C). Margarine must melt below this temperature.

(b) (i) It decreases. (ii) Zero.

(c) Amount of I_2 = (25/1000) x 0.1 mol = 0.0025 mol. $M_m(I_2)$ = 254 g mol^{-1}.
Mass of I_2 = 0.0025 x 254 g = 0.635 g

(d) (i) Amount of $Na_2S_2O_3$ = (23.4/1000) x 0.1 mol = 0.00234 mol. $M_m(Na_2S_2O_3)$ = 158 g mol^{-1}. Mass of $Na_2S_2O_3$ = 0.370 g.
(ii) 2 x 158 g $Na_2S_2O_3$ ≡ 254 g I_2. Hence 0.370 g $Na_2S_2O_3$ ≡ 0.297 g I_2.
(iii) 0.635 – 0.297 g I_2 = 0.338 g I_2.
(iv) 0.56 g fat ≡ 0.338 g I_2. Therefore 100 g fat ≡ 60.34 g I_2. I.V. = 60.

9. (a) $H_3C-CH_2-C(=O)-O-H + H-CH_2-O-H \longrightarrow H_3C-CH_2-C(=O)-O-CH_3 + H_2O$

(b) Methyl propanoate.

(c) $R-C(=O)-O-R'$

In esters, R′ must be a carbon-containing group of some kind; it cannot be H (which makes the molecule an acid. Note, however, that R is H when the parent acid is methanoic acid.)

10. A: propyl ethanoate; B: ethanol; C: methanoic acid.

11. (a) $CH_3-C(=O)-O-C_3H_7$ (b) $C_2H_5-C(=O)-O-C_2H_5$ (c) $H-C(=O)-O-C_4H_9$

12. $H_3C-C(=O)-OH + H_3C-{}^{18}O-H \longrightarrow H_3C-C(=O)-{}^{18}O-CH_3 + H-O-H$

Unit 4 Answers

13. Forward reaction: condensation or esterification; reverse reaction: hydrolysis.

14. (a)

Ester	Short structural formula of: X	Y	Acid	Alcohol
1.	C_6H_5–OH (benzene ring with OH)	CH_3	C_6H_4(OH)COOH (benzene ring with COOH and OH)	CH_3OH
2.	3 x $C_{15}H_{31}$	CH_2CHCH_2	3 x $C_{15}H_{31}COOH$	$CH_2OHCHOHCH_2OH$
3.	H	C_2H_5	HCOOH	C_2H_5OH
4.	CH_3	C_6H_5–COOH (benzene ring with COOH)	CH_3COOH	C_6H_4(OH)COOH (benzene ring with COOH and OH)

(b)

```
    H
    |
H-C-O-H
    |
H-C-O-H
    |
H-C-O-H
    |
    H
```

(c) 1: phenol;
2:triglyceride;
3: aldehyde;
4: acid.

15. (a) 168 ± 2 °C

(b)

```
  H H H     O
  | | |   //      H H
H-C-C-C-C         | |
  | | |   \ O-C-C-H
  H H H           | |
                  H H
```

16. (a) (Alkaline) hydrolysis.

(b) (i), (ii) 1: $HCOONa$, $C_5H_{11}OH$; sodium methanoate, pentanol. 2: C_3H_7COONa, CH_3OH; sodium butanoate, methanol. 3: C_4H_9COONa, C_2H_5OH ; sodium pentanoate, ethanol. 4: $C_5H_{11}COONa$, C_3H_7OH; sodium hexanoate, propanol. 5: CH_3COONa, $C_6H_{13}OH$; sodium ethanoate, hexanol. 6: C_2H_5COONa, C_4H_9OH; sodium propanoate, butanol.

17. (a) Linear: lengthwise, one-dimensional, without branching; thermoplastic: softens when heated; synthetic: artificially made, not made by a natural process; monomers: single units that join to make a polymer.

(b) Benzene-1,4-dicarboxylic acid; ethane-1,2-diol.

(c) (i), (ii) and (iii).

```
  O            O                         O             O
  ||           ||                        ||            ||
-C-(C6H4)-[C-O]-CH2CH2-O-C-(C6H4)-C-
```

or

```
                     O             O
                     ||            ||
-O-CH2CH2-[O-C]-(C6H4)-C-O-CH2CH2-O-
```

(d) Crude oil.

(e) The carboxylic groups in terephthallic acid are in the opposed 1,4 positions. These give narrow polymer chains that pack closely: good fibres. In phthallic acid, however, the carboxylic groups are in adjacent 1,2 positions. The bulky benzene rings stick out of the chains preventing close packing: poor fibres.

18. (a) Three-dimensional polymer: giant molecule in which bonding extends in all directions; thermosetting: cannot be softened once set; cross-linking: bonds between linear molecules; addition: unsaturated molecules join; glass fibre reinforcement: glass fibres embedded in liquid plastic as it sets.

(b) Phenylethene.

(c) (i), (ii)

19. (a) (i) The Haber process; (ii) $N_2 + O_2 \rightarrow 2NO$; (iii) Some nitrifying bacteria live in nodules on the roots of plants like clover and alfalfa. These bacteria work in symbiosis with the plant to convert elemental nitrogen into compound form.
  (b) Nitrogen fixation.
  (c) (i) Plants synthesise protein from nitrogen compounds in the soil.
  (ii) Animals make their own kind of protein out of plant and animal protein; animals cannot synthesise protein from non-protein nitrogen compounds.
20. (a) $(NH_4)_2SO_4 + 2NaOH \rightarrow 2NH_3 + Na_2SO_4 + 2H_2O$.
  (b) It prevents any of the contents of the flask being carried over during the distillation.
  (c) The acid (excess) that has *not* been consumed by the ammonia is determined by titration. The acid consumed is the difference between this excess and the total acid added. This kind of titrimetric analysis is called an indirect titration or back titration. (In a direct titration, the ammonia is distilled over into water and titrated directly.)
  (d) (i) (1/6.25) x 100% = 16.0%; (ii) % nitrogen = (1/6.38) x 100% = 15.7%: poorer.
  (e) A reference solution of any alkali, e.g., 0.1 mol L^{-1} NaOH.
  (f) (50 − 38) mL = 12 mL used up.
  (g) (i) $2NH_3 + H_2SO_4 \rightarrow (NH_4)_2SO_4$

  $n = 2$ mol; $n = 1$ mol

  $m = 34$ g; $V = 10\,000$ mL of 0.1 mol L^{-1}

  10 000 mL of the acid ≡ 34 g NH_3

  12 mL of the acid ≡ 0.041 g NH_3, which is the mass of ammonia absorbed.

  (ii) 17 g ammonia contains 14 g nitrogen. 0.041 g contains 0.034 g nitrogen.
  (iii) Mass of protein in the sample (1.72 g) of flour is 0.034 x 6.25 g = 0.213 g. Percentage protein in (this) flour = (0.213/1.72) x 100% = 12.4%.
21. (a) Amino acids: these have the short formula* $NH_2{-}CHX{-}COOH$ where X is a variable group; proteins: naturally occurring polypeptides of high molecular mass†.
  *Proline has a different structure.
  †Some proteins consist of two or more crosslinked polypeptides.
  (b) (i)
```
    H H O H H O
    | | ‖ | | ‖
  H-N-C-C-N-C-C-O-H
      |     |
      H     H
```
  (ii)
```
    H H O H H O H H O
    | | ‖ | | ‖ | | ‖
  H-N-C-C-N-C-C-N-C-C-O-H
      |     |     |
      H     H     H
```
  (iii)
```
   H H O H H O H H O
   | | ‖ | | ‖ | | ‖
  -N-C-C-N-C-C-N-C-C-
     |     |     |
     H     H     H
```
  (iv)
```
   O H
   ‖ |
  -C-N-
    ↑
```
  (c) Condensation polymerisation.
22. NH_2CH_2COOH.
23. (i) $NH_2CH(CH_3)COOH + NaOH \rightarrow NH_2CH(CH_3)COONa + H_2O$;
  (ii) $HCl + NH_2CH(CH_3)COOH \rightarrow [NH_3CH(CH_3)COOH]Cl + H_2O$;
  (iii) $NH_2CH(CH_3)COOH + CH_3OH \rightarrow NH_2CH(CH_3)COOCH_3 + H_2O$.
24. (a) The hydrolysate contains 4 (glycine). It also contains 3 and/or 5 (valine and/or tryptophan).
  (b) (i) Because 3 and 5 are not separated by solvent 1. We therefore do not know if the hydrolysate contains 3 only, 5 only, or both.
  (ii) As a confirmatory check on 4. (c) Valine and glycine.

25. Gly-Ala-Cys: NH_2CH_2–**CONHCH(CH_3)CONH**–CH(CH_2SH)COOH;
Cys-Ala-Gly: NH_2(CH_2SH)–**CONHCH(CH_3)CONH**–CH_2COOH.
The middle portions of each tripeptide (in bold) are the same, but the ends are different.

26. (a) An enzyme.
(b) Lock and key mechanism: an enzyme molecule has a unique shape that fits the substrate molecule(s) only. This makes the enzyme selective.

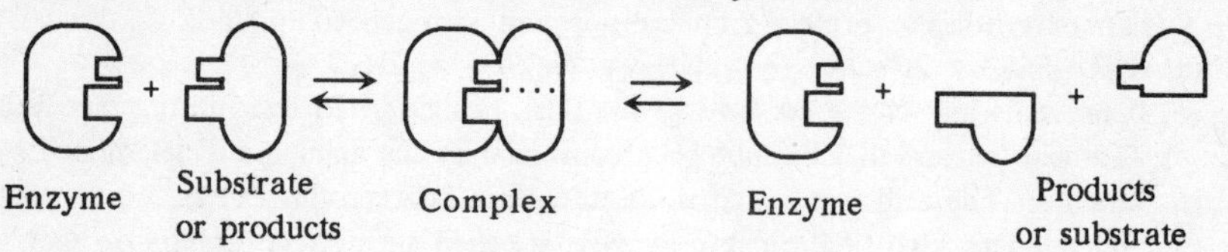

(c) They pass into the bloodstream and are rebuilt into protein as required by the body. The nitrogen of excess amino acids is converted to urea and expelled while the C/H/O may be stored as fat.
(d) Phenylalanine is described as an essential amino acid (as are eight or nine others.)
(e) (i), (ii) CH_2; H, O; ····–N–C–C–····; H; Cleaved bond
(f) Aromatic side chains.

27. t/s: 0, 5, 10; T/°C: 0 10 20 30 40 50 60

The optimum temperature is the temperature at which the rate of reaction is a maximum and the time taken a minimum. From the graph, optimum temperature is 35 °C.

28. For 20 amino acids there are 20 x 19 x 18 x ... x 3 x 2 x 1 = 2.43 x 10^{18} possible arrangements. (The sequence 20 x ... x 1 is called factorial 20 and is written as 20! Scientific calculators have a factorial key.)

29. (a)

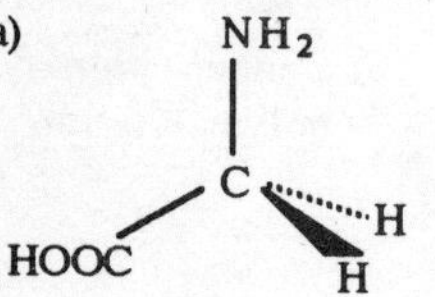

Glycine does not have four different groups; two are the same (H atoms). A mirror image structure is not possible. (A box of atomic models will help you to understand this idea if you find it difficult.)

(b)

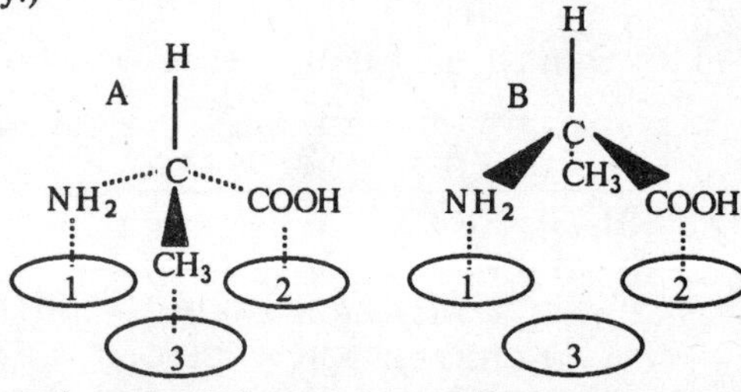

A is the natural alanine: the amino, carboxyl and methyl groups fit exactly onto their respective receptors. This is not the case with B: only two of the receptors can be engaged at any one time.

30. (a)

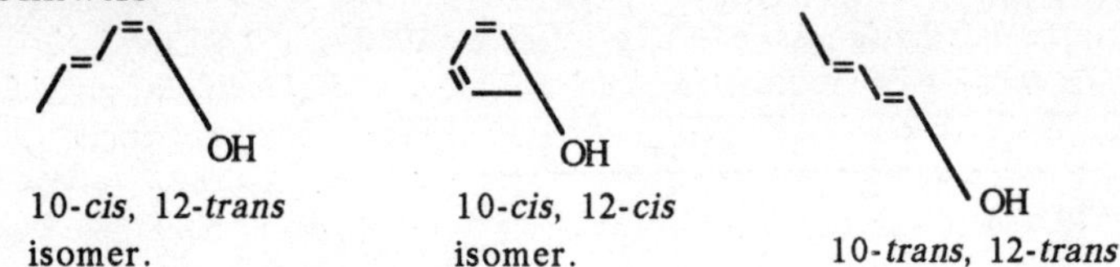

10-*cis*, 12-*trans* isomer. 10-*cis*, 12-*cis* isomer. 10-*trans*, 12-*trans* isomer.

(b) The 10-*trans*, 12-*cis* isomer.

(c) Concentration of 10-*trans*, 12-*cis* isomer is 10^{-18} g cm^{-3}.
Molar mass = 238 g mol^{-1}.
Since 238 g consists of 6 x 10^{23} molecules, 10^{-18} g consists of 2521 molecules.
No. of molecules cm^{-3} = 2.52 x 10^3 cm^{-3}.

(d) By a factor of $10^{18}/10^5$ = 10^{13} = 10 000 000 000 000 = 10 million million.

31. (a) Denaturation: the loss of biological nature and activity; globular proteins: proteins that are soluble in aqueous solutions of one kind or another (fibrous proteins like silk fibroin are insoluble); colloidal solutions (dispersions): the solute particles are too large to go into true solution but not so large as to give a suspension. (The stability of colloidal solutions depends on the repulsion between particles due to like charges on their surfaces.)

(b) (i) A highly organised structure becomes a disorganised structure.
(ii) Peptide bonds are not broken; the disorganised structure consists of the same polypeptide(s) as the organised structure.

(c) Select two from: heat, altered pH, vigorous shaking or whisking, organic liquids, heavy metal ions, e.g. Pb^{2+}.

32. (a) Proteolytic: enzyme-splitting; coagulation: clotting; supernatant: the clear liquid that remains after a precipitate has settled; optimum pH: the pH at which an enzyme works best.

(b) pH = 6.6 (from tube 3).

(c) 38 °C would be best but anything between about 30 °C and 40 °C would do.

(d) To keep the volume variable constant. (Five drops in total are added to each tube.)

(e) (i)

t/s: 0, 50, 100
pH: 3 4 5 6 7 8 9

(ii) Extrapolation (= continuing the curve; the dotted line) gives a minimum time (maximum rate) at about pH 3, which is therefore the optimum.
(iii) Extrapolation is equivalent to guessing and so the optimum pH is not known for sure. (pHs less than about 5.0 cannot be used in this experiment because milk curdles at these values in the absence of rennin.)

(f) Cheesemaking, junket making.

Unit 5

1. (a) (i) The Noble Gases. (ii) Unreactive: reacts sometimes but not often. Inert: does not react at all. (iii) Helium, neon or argon.

(b) 24 800 mL xenon has a mass of 131 g. Hence, 30 mL ..... 0.158 g.
Therefore, 0.249 g of the fluoride contains 0.158 g xenon and 0.091 g fluorine.

(c)

Elements	Xenon		Fluorine
Mass ratio	0.158	:	0.091
A_r	131		19
Rough ratio of atoms	0.158/131	:	0.091/19
	= 0.0012	:	0.0048
Whole number ratio	= 1	:	4
Empirical formula	= XeF_4		

(d) Molecular mass.
(e) $XeF_4 \rightarrow Xe + 2F_2$ (Note that xenon, unlike fluorine, is monatomic.)

2. (a) (i) Fluorine: 85 K; chlorine: 238 K. (ii) Chlorine.
(b) (i) The bonding within a fluorine molecule is a single covalent bond. This consists of two shared electrons, one from each fluorine atom.
(ii) The bonding between molecules is van der Waals attraction. This is due to transient changes in electron density within molecules. These changes create positive and negative regions on the molecules, which then attract.
(c) (i) Van der Waals attraction increases with molecular size, and so gases give way to liquids, which give way to solids.
(ii) Halogen atoms each have 7 outer electrons and achieve stability by increasing this number to 8. The force with which this is done decreases as the atom becomes bigger because the extra electron is farther from the positive nucleus.

3. (a) H:H, (b) H:F:, (c) H:Cl:, (d) H:Br:, (e) :F:F:, (f) :Br:Br:, (g) :I:I:,
(h) H:C:H (with H above and H below C), (i) H:Si:H (with H above and H below Si), (j) :Cl:C:Cl: (with :Cl: above and :Cl: below C), (k) :Br:Si:Br: (with :Br: above and :Br: below Si), (l) :N:H (with H above and H below N), (m) :P:H (with H above and H below P),
(n) :P:I: (with :I: above and :I: below P), (o) H:O:H, (p) H:S:H, (q) :O::O:, (r) :N:::N:, (s) :O::C::O:,
(t) H:C:C:H (with H H above and H H below), (u) H:C::C:H (with H H above).

4. (a) An alloy. (b) Copper atoms all have the same size (117 pm) and, because of this, can exchange lattice positions: pure copper is soft. The presence of tin atoms, which are larger (140 pm), makes exchange more difficult and so the alloy is harder.
5. (a) For a given row, main group elements (elements in groups 1–8) have very different chemical properties. But transition elements (elements in the ten groups from Sc to Zn) have similar chemical properties.
(b) Potassium: 0.86 g cm^{-3}; copper: 8.92 g cm^{-3}.
(c) (i) Packing: copper atoms are more closely packed than potassium atoms.
Radii: copper atoms are smaller. (ii) Copper atoms have a greater mass (64 u) than potassium atoms (39 u).
6. (a) A. (b) I. (c) E. (d) F. (e) D. (f) C and H. (g) B and I. (h) E and G. (i) E and G.
7. (a) A: C (graphite). B: Mg. C: S (S_8). D: Al. E: N_2. (Note triple bond.) F: Li.

(b) (i) Greasy: the bonding between the planes of C atoms is weak van der Waals attraction, which allows the planes to slip past each other.
(ii) Good conductor: each carbon atom uses only three of its four valency electrons for covalent bonds. The fourth electron is mobile and carries charge as in a metal.

(c)

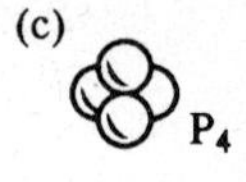

8. (a) (i) Covalent radius. (ii) Bond length. (b) $c = b - (a/2) = 77.5$ pm.
9. (a) (i) 5.62; (ii) 1.89.
   (b) The gain of one electron by an atom that has only one electron (H) causes a much greater increase in size than the gain of one electron by an atom (F) that has nine.
10. (a) (i) Increasing nuclear charge. (These ions each have the same two electrons.)
    (ii) Additional electron shell. (C^{4+} is 2 while N^{3-} is 2.8.)
    (iii) Increasing nuclear charge. (These ions each have the same 2.8 configuration.)
    (b) In principle, a line graph is a plot of a continuous variable (y) against a continuous variable (x). Example: temperature against time. Ideally, a bar graph is a plot of a discrete variable against a discrete variable. Example: number of isotopes against atomic number. Spike graphs are usually a plot of a continuous variable against a discrete variable. Example: ionic radii against named ion (as in this question).
11. (a) Neighbours differ by one proton.
    (b) Same electron configuration (2.8).
    (c) Equal radii.
    (d) Increasing nuclear charge pulls in electrons more and more, reading from left to right.
    (e) (i) Ne does not form an ion. (ii) About (136 + 95)/2 pm = 116 pm.
12. (a) (i) 1.602×10^{-19} J x 5 = 8.01×10^{-19} J; (ii) 1.43×10^{-18} J.
    (b) Kinetic energy equal to the (first) ionisation energy of helium.
    (c) (i) $He(g) \rightarrow He^{+}(g) + e^{-}$; (ii) 1.602×10^{-19} J x 24.6 = 3.94×10^{-18} J;
    (iii) 3.94×10^{-18} J x 6×10^{23} = 2364 000 J = 2364 kJ.
13. (a) (i) R; (ii) Second ionisation energy is much greater than first. R must therefore have only one outer electron and is in Group 1.
    (b) S. (1260 + 2310 + 3840 kJ = 7410 kJ.)
14. (a) (i) H (1)-He (2). The He atom is smaller. Its electron is more difficult to remove because it is nearer the nucleus. Also, the complete duplet in helium is stable.
    (ii) He (2)-Li (2.1). The Li atom is bigger. Its electron is removed more easily because it is farther from the nucleus. Also, the complete inner duplet in Li screens the outer electron from the nucleus and makes its removal easier still.
    (iii) Li (2.1)-Be (2.2). The Be atom is smaller. Its electron is more difficult to remove because it is nearer the nucleus.
    (b) He^{+} (1)-Li^{+} (2). The removal of an electron from Li^{+} requires breaking into the stable duplet. Hence the second ionisation energy of Li is very high.
    (c) The hydrogen atom has only one electron.
    (d) The removal of a negative electron from an already positive ion requires more energy than does its removal from a neutral atom.
    (e) Be.
    (f) Be (2.2)-Ne (2.8). Nuclear charge increases and atom size decreases. Both factors

make electron removal more difficult and cause the ionisation energies to increase.

15. 2.8.8.1. (K)
16. (a) Group 1.
    (b) Malleable: can be beaten into shape; ductile: can be drawn out and stretched.
    (c) (i) All have one outermost electron; (ii) The inner shells vary from 2 (Li) to 2.8.18.18.8 (Cs).
    (d) $Li + H_2O \rightarrow LiOH + \frac{1}{2}H_2$; $Na + H_2O \rightarrow NaOH + \frac{1}{2}H_2$; $K + H_2O \rightarrow KOH + \frac{1}{2}H_2$; $Rb + H_2O \rightarrow RbOH + \frac{1}{2}H_2$; $Cs + H_2O \rightarrow CsOH + \frac{1}{2}H_2$. Select two.
    (e) Reactions depend on the loss of the lone outer electron. Its loss becomes easier down the group as atom size increases. Nuclear charge also increases, which should make loss more difficult. But the outer electron is screened from the nucleus by the inner electron shells.
17. (a) H_2 NH_3 HF LiI NaCl CsF.
    (b) Polar covalent molecule. The symbols δ+ and δ– represent partial positive and negative charges. (HCl in the gaseous state is about 5% ionic, 95% covalent.)
    (c) 92-95%.
18. (i) NaH; (ii) Blue ; (iii) CH_4; (iv) No effect; (v) NH_3; (vi) Blue; (vii) HCl; (viii) Red.
19. (a) Difference in electronegativity of the element and that of hydrogen. (Electronegativity of an element is its pull on bonding electrons.) Most Group 1 and 2 elements are so much less electronegative than hydrogen that they form ionic hydrides in which the hydrogen is present as a negative hydride ion. The remaining elements form polar covalent or covalent hydrides.
    (b) (i) PH_3 (though SiH_4 might also be accepted). (ii) Polarity of the hydrides across the row swings from + – for NaH to δ–δ+ for ClH with the non-polar (= 100% covalent) P–H bond in the middle of the row.
20. (a) (i) Direct synthesis: combination between the metal and hydrogen. Decomposition: the breaking down of a compound into simpler parts. Spontaneous ignition: catching fire of its own accord (without help).
    (ii)

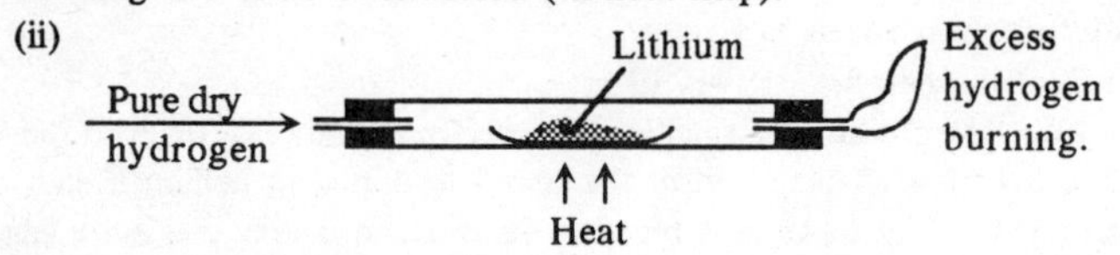

    (b) Decreases. (c) $2CsH + O_2 \rightarrow Cs_2O + H_2O$.
21. (a) X: $Li^+ + e^- \rightarrow Li$. Y: $H^- \rightarrow \frac{1}{2}H_2 + e^-$.
    (b) Hydrogen is usually evolved from the cathode (negative electrode), not the anode (positive electrode) as here.
    (c) Because lithium hydride reacts with water: $LiH + H_2O \rightarrow LiOH + H_2$.
22. (a) (i) The increase in nuclear charge pulls in the outer shell.
    (ii) The increase in electron shells.
    (b) Neon does not react and so has no covalent radius.
    (c) (i) The loss of the outer electron leaves a smaller particle: the outer shell is lost.
    (ii) The gain of an outer electron creates a larger particle: the outer shell is larger.
23. (a) Ionic lattice.

(b) (i) 0.81. (ii) CsCl structure. (iii) NaCl structure. Each ion has six oppositely charged ions around it. (iv) No. (Irrespective of radius ratio values, the only alkali halides with 8 co-ordination are CsCl, CsBr and CsI.)

24. (a)

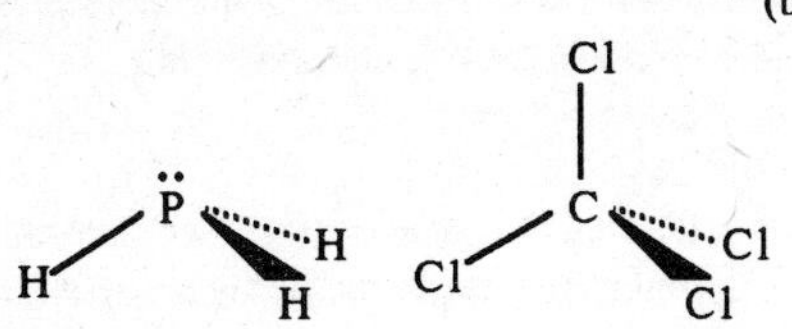

(b) PH_3: The non-bonding electron pair on the phosphorus atom makes the top of the molecule negative and the rest positive. CCl_4: The polar bonds cancel because the molecule is symmetrical.

25. (i) The O–H bonds are polar and the molecule is non-linear (bent):

δ− :Ö —H, H δ+

(ii) The negative ends of water molecules are attracted to a positive surface and vice versa:

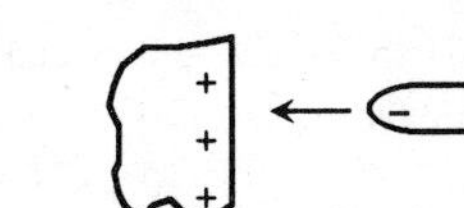

(iii) If like charges oppose, the polar molecule rotates so that unlike charges attract.

(iv) A charged surface induces a slight opposite charge on molecules and attracts them.

26. (a) 1: polar; 2: non-polar; 3: polar.

(b) A charged rod is held near a stream of the liquid under test. Strong deflection means the liquid is polar. Slight deflection means it is non-polar.

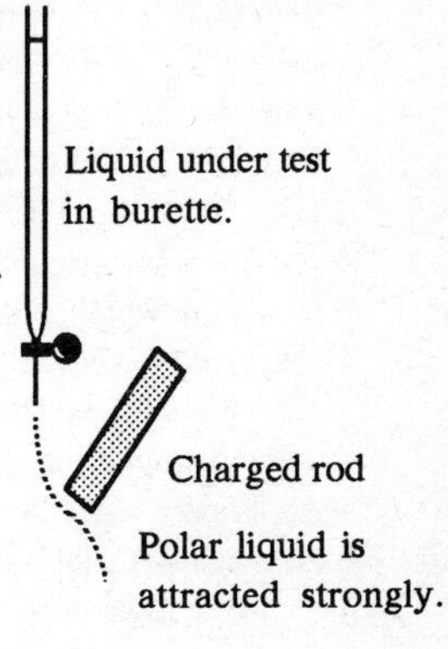

(c) (i) Liquid 2.

(ii) It is non-polar: non-polar molecules are less attracted to each other than polar molecules. The b.p. of the non-polar liquid should therefore be lower.

27.

In its rest position, the molecule has this shape and is non-polar.

Near a charged surface, rotation about the C–C bond occurs, the molecule adopts this shape and is polar.

28. (a) Van der Waals attraction increases with molecular size and b.p. therefore rises.

(b) $CH(CH_3)_3$ (H–C–H group attached to the central C of H–C–C–C–H, all other positions H)

(c) (i) At a lower temperature.

(ii) Branched molecules are more compact than their straight-chain isomers. Van der Waals attraction between the branched molecules is therefore weaker and b.p. lower.

(iii) Y is the most branched of the C_5 isomers:

CH_3–C(CH_3)$_2$–CH_3

(iv) Only one structural formula is possible for C_3H_8 and so there are no isomers.

(d) Methanol, CH_3OH.

(e) Two kinds of between-molecule attraction operate in alcohols: van der Waals and hydrogen bonding. In alkanes, only van der Waals forces hold the molecules together. Alcohols therefore have higher b.p.s for the same molecular mass.

29. Hydrogen chloride is a covalent gas that is ionised by water,

$$HCl(g) \longrightarrow H^+(aq) + Cl^-(aq),$$

giving a conducting solution. The gas dissolves in benzene without ionisation.

30. Water molecules are extensively hydrogen bonded into large (variable) groups. This reduces their mobility and increases the viscosity of the liquid.

31. (a) Viscosity: the ability of a liquid to flow. (Syrup has high viscosity; it is viscous. Petrol has low viscosity; it is mobile.)

(b) Ether.

(c) At a mole ratio of 1:1, i.e., when the amount-percentage of each component is 50%.

(d) Hydrogen bonding creates dimers:

$$\delta- \left\{ \begin{matrix} Cl \diagdown \\ Cl-\overset{\delta+}{C}-H \cdots \overset{\delta-}{O} \\ Cl \diagup \end{matrix} \right. \quad \begin{matrix} \diagup C_2H_5 \\ \\ \diagdown C_2H_5 \end{matrix}$$

(The combined effect of three chlorine atoms makes the hydrogen atom $\delta+$ enough to form a hydrogen bond.)

32. (a)

(i)	I_2	254 u molecule^{-1}.	Should be solid.	Correct prediction.
(ii)	S_8	256 u molecule^{-1}.	Should be solid.	Correct prediction.
(iii)	CO_2	44 u molecule^{-1}.	Should be gas.	Correct prediction.
(iv)	SiO_2	60 u molecule^{-1}.	Should be liquid.	Incorrect prediction.
(v)	H_2O	18 u molecule^{-1}.	Should be gas.	Incorrect prediction.
(vi)	H_2S	34 u molecule^{-1}.	Should be gas.	Correct prediction.

(b) (iv) Silicon dioxide (quartz, sand, etc.) is not composed of discrete SiO_2 molecules; it is a network solid (giant molecule). The value 60 u molecule^{-1} makes the wrong assumption that SiO_2 is discrete like CO_2.

(v) Water molecules are associated by hydrogen bonding so that the effective molecular mass is much greater.

33. (a) A: silicon dioxide (SiO_2); B: ice (H_2O).

(b) A: covalent macromolecular lattice; B: discrete molecular lattice.

(c) A has a high m.p. because strong covalent bonds have to be broken when it melts. B has a low m.p. because weak hydrogen bonds need only be broken for it to melt.

(d) (i) Silicon carbide (carborundum), diamond, boron nitride. (ii) They are very hard.

34. Ice is extensively hydrogen bonded and so has an open structure of low density. It floats on water. Hydrogen bonding in water decreases over the range 0 °C to 100 °C. This would make the water denser. But the increase in molecular motion that goes with increase in temperature would make it lighter. The two opposing effects give a greatest density at 4 °C.

35. (a) Anomalous: unusual, unexpected.

(b) It is higher than would be expected for a substance composed of small HF molecules. Extrapolation of the graph would give an expected b.p. of about −100 °C.

(c) (i) Hydrogen bonding: the polar molecules become associated by electrostatic attraction with $\delta+$ hydrogen atoms between $\delta-$ fluorine atoms.

(ii) 40 u: H–F···H–F; 60 u: H–F···H–F···H–F.

36. (a) (i) CH_4:16 u; NH_3:17 u.
    (b) Van der Waals attraction between molecules of same mass is about the same.
    (c) Ammonia molecules are held together by hydrogen bonds: $H_3N \cdots H{-}NH_2$.
37. (a) Associated: joined together, usually weakly; dimer: a particle made from two simpler particles (monomers).
    (b) The hydrogen bond.
    (c) (i) By electrostatic attraction between δ– oxygen atoms on each side of a δ+ hydrogen atom.
    (ii) The molecule must contain hydrogen; it must contain an electronegative element, e.g., nitrogen, oxygen or fluorine; there must be a lone pair of electrons on the electronegative atom.
38. (a) Carbon (diamond), silicon carbide, silicon dioxide, fluorine.
    (b) Hydrogen chloride, hydrogen oxide.
    (c) Hydrogen oxide. (Possibly hydrogen chloride though the effect must be slight.)
    (d) Carbon (graphite).
    (e) Gold.
    (f) Caesium fluoride.
    (g) Neon.
    (h) Carbon (graphite).
    (i) Hydrogen chloride, fluorine.
39. (a) Na_3AlO_3 or $NaAlO_2$; (b) Na_2ZnO_2; (c) Na_2PbO_3; (d) Na_2PbO_2; (e) Na_2VO_3; (f) Na_3AsO_3 or $NaAsO_2$; (g) Na_2SnO_2; (h) Na_3SbO_3 or $NaSbO_2$. (These formulae are obtained by writing the oxide as hydroxide and replacing the hydogen atoms with sodium: e.g., $Al_2O_3 \rightarrow Al(OH)_3 \rightarrow H_3AlO_3/HAlO_2 \rightarrow Na_3AlO_3/NaAlO_2$.)
40. (a) The trend is a result of differences in electronegativity. In the metal hydroxide, the least electronegative element is the metal. Dissociation therefore gives metal$^+$(aq) and OH^-(aq). The solution is alkaline. In the non-metal hydroxide, the least electronegative element is hydrogen. Ionisation gives H^+(aq) and an anion. The solution is acidic. (Examples: $CsOH \rightarrow Cs^+ + OH^-$; $SO_2(OH)_2 = H_2SO_4 \rightarrow 2H^+ + SO_4^{2-}$.)
    (b) An oxide that reacts as a base with an acid and as an acid with a base.
    (c) Al_2O_3. As a base: $6HCl + Al_2O_3 \longrightarrow 2AlCl_3 + 3H_2O$;
    As an acid: $Al_2O_3 + 2NaOH \longrightarrow 2NaAlO_2 + H_2O$.
    Or ZnO. As a base: $2HCl + ZnO \longrightarrow ZnCl_2 + H_2O$;
    As an acid: $ZnO + 2NaOH \longrightarrow Na_2ZnO_2 + H_2O$.
    (d) Any amino acid, e.g., glycine. (It is amphoteric for a different reason: it can react as a base because of the NH_2– group and as an acid because of the –COOH group.)
41. The bonding in lithium iodide is about 50% ionic, 50% covalent. In caesium fluoride the bonding is ionic with very little covalency.
42. (a) The oxygen atom of a water molecule bonds to a positive ion. The hydrogen atoms bond to a negative ion. Reason: the water molecule is a polar molecule in which the oxygen is δ– and the hydrogen δ+.
    (b) Small ions are more heavily hydrated than large ions and so cannot move as quickly.
    (c) Lattice energy is absorbed to remove the ions from their lattice positions. Hydration

energy is evolved as water molecules become attached to the ions. The difference between these terms is the heat of solution.

43. (a) Dissociation. (This is not ionisation because the ions already exist.)
(b) Hydrolysis. (c) Hydration. (d) Ionisation. (e) Ionisation.

44. Hydrolysis of PCl_3: Water splits the PCl_3 and is itself split.

$$PCl_3(l) + 3H_2O(l) \longrightarrow H_3PO_3(aq) + 3HCl(g)$$

Dissociation of NaCl: Water allows the ions to move apart.

$$Na^+Cl^-(s) + \text{water} \longrightarrow Na^+(aq) + Cl^-(aq)$$

Ionisation of HCl: Water creates ions where none had existed.

$$HCl(g) + \text{water} \longrightarrow H^+(aq) + Cl^-(aq)$$

Hydration of copper sulphate: Water molecules become bonded to the ions in the solid.

$$\underset{\text{(anhydrous solid)}}{CuSO_4(s)} + 5H_2O(l) \longrightarrow \underset{\text{(hydrated solid)}}{CuSO_4.5H_2O(s)}$$

45. (a) (i) Fluorine; (ii) Iodine; (iii) Iodide ion; (iv) Fluoride ion.
(b) Fluorine—chlorine—bromine—iodine.

46. (a) Bromine vapour.
(b) (i) $Cl_2 + 2e^- \longrightarrow 2Cl^-$; $2Br^- \longrightarrow Br_2 + 2e^-$.
(ii) $Cl_2 \rightarrow 2Cl^-$ is a reduction; $2Br^- \rightarrow Br_2$ is an oxidation.
(c) (i) $Cl_2 + 2Br^- \longrightarrow 2Cl^- + Br_2$.
(ii) Displacement reaction (chlorine displaces bromine). It is also a redox reaction.
(d) (i) Iodine.
(ii) Starch solution — gives a blue/black colour.
(e) $Br_2 + 2I^- \longrightarrow 2Br^- + I_2$.
(f) It demonstrates that neither chlorine nor bromine survive being passed through these solutions. (Chlorine, and bromine to a lesser extent, bleach litmus paper.)

Unit 6

1. (a) It allows efficient heat transfer from hot gases to surrounding water.
(b) Calorimeter would begin to lose heat to surroundings.
(c) $CH_3OH + 1^1/_2O_2 \longrightarrow CO_2 + 2H_2O$.
(d) (i) $cm\Delta T$ = 4.18 kJ K^{-1} kg^{-1} x 0.8 kg x 2.6 K = 8.69 kJ. (ii) 0.412 g CH_3OH gives 8.69 kJ heat. So 32 g (mass of 1 mol) give 675 kJ: ΔH = -675 kJ mol^{-1}.
(e) (i) -715 kJ mol^{-1}. (ii) Various: heat loss from calorimeter; incomplete transfer of heat through coil; incomplete combustion.

2. (a) To ensure complete combustion of the food — air is only $^1/_5$th oxygen.
(b) The filament would glow and ignite and set fire to the food.
(c) (i) The heat generated by the filament should be subtracted from that of the food + filament. (ii) Determine the heat produced by the filament only.
(d) Calorific value.

3. $\Delta T = Q/cm$ from which ΔT = 38.6 °C.

4. ΔH = +50 kJ mol^{-1}.

5. (a) 20.3 °C.
(b) (i) At 5 s; (ii) 9-10 s.
(c) ΔT = (27.2 – 20.3) °C = 6.9 °C. See graph on following page.
(d) The temperature rise would be the same for 1 L of acid mixed with 1 L of alkali, for which m = 2 kg. $Q = cm\Delta T$ = 57.7 kJ. ΔH = -57.7 kJ mol^{-1}.

6. (a) 7 °C. (b) 14 °C. (c) 4.7 °C. (d) 3.5 °C. (e) 3.5 °C. (f) 14 °C.

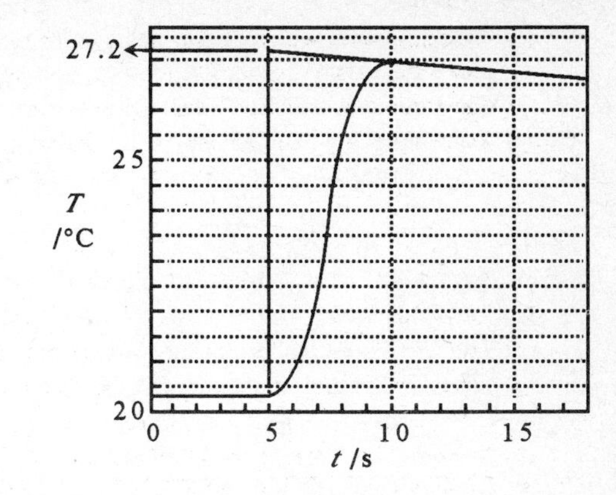

7. For 0.64 g NH_4NO_3, $Q = cm\Delta T$ = 0.21 kJ.
   $M_m(NH_4NO_3)$ = 80 g mol^{-1}.
   For 80 g NH_4NO_3, Q = 26.3 kJ.
   Hence $\Delta H_{soln.}$ = +26.3 kJ mol^{-1}.
8. Ion size and ion charge. The bigger the ion, the smaller the heat evolved, e.g., size increases from Li^+ to Na^+ and evolved heat decreases. The greater the charge, the greater the heat evolved, e.g., Li^+ and Mg^{2+} are about the same size but hydration of Mg^{2+} evolves more heat.
9. (a) (i) N_2O. (ii) Exothermic compound.
   (b) (i) $C(s) + O_2(g) \rightarrow CO_2(g)$ ... eqn. 1, $N_2(g) + \frac{1}{2}O_2(g) \rightarrow N_2O(g)$ ... eqn. 2;
   (ii) $C(s) + 2N_2O(g) \rightarrow CO_2(g) + 2N_2(g)$.
   (c)

	Equation	ΔH/kJ mol^{-1}
Eqn. 1:	$C(s) + O_2(g) \rightarrow CO_2(g)$	-394
Double and reverse eqn. 2:	$2N_2O(g) \rightarrow 2N_2(g) + O_2(g)$	-148
Add:	$C(s) + 2N_2O(g) \rightarrow CO_2(g) + 2N_2(g)$	-542

   ΔH = -542 kJ mol^{-1}.
   (d) They have zero enthalpies of formation.
   (e) (i) (ii)

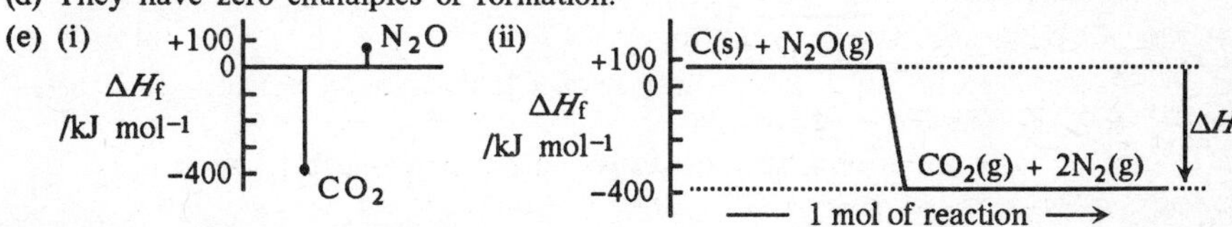

10. (a) $H_2(g) + \frac{1}{2}O_2(g) \rightarrow H_2O(l)$.
    (b) $C(s) + 2H_2(g) \rightarrow CH_4(g)$. (c) $3C(s) + 3H_2(g) \rightarrow C_3H_6(g)$.
    (d) $N_2(g) + 2H_2(g) + 1\frac{1}{2}O_2(g) \rightarrow NH_4NO_3(s)$.
    (e) $2C(s) + 3H_2(g) + \frac{1}{2}O_2(g) \rightarrow C_2H_5OH(l)$.
    (f) $6C(s) + 6H_2(g) + 3O_2(g) \rightarrow C_6H_{12}O_6(s)$.
    (g) $16C(s) + 9H_2(g) + 2O_2(g) + N_2(g) + S(s) \rightarrow C_{16}H_{18}O_4N_2S(s)$ (penicillin).
11. (a) Because the heats of combustion increase by equal CH_2 increments.
    (b) Extend the line and read off the graph. ΔH (butanol) ≈ 2600 kJ mol^{-1}.
    (c) The origin corresponds to the heat of combustion of '$CH_3OH - CH_2$' = H_2O, which is zero.
12. (a) ΔH_1: lattice energy; ΔH_2: hydration; ΔH_3: solution.
    (b) (i) $\Delta H_1 + \Delta H_2 = \Delta H_3$. (ii) ΔH_1 = +635.5 kJ mol^{-1}. (iii) Temperature falls.
    (c) $-322 + x = -615$ ... $x = -293$. ΔH for $I^-(g) \rightarrow I^-(aq)$ is -293 kJ mol^{-1}.
13. x = +1022; y = -902; z = -18.

14. (a) 1: compounds with positive heats of formation are less stable. 2: RDX is a solid that gives gases on decomposition. 3: decomposition is highly exothermic.
(b) The rate at which RDX decomposes is extremely rapid (detonation).

15. (a)

Bond breaking	ΔH/kJ mol^{-1}	Bond making	ΔH/kJ mol^{-1}
C(s→g)	+715	2 x C=O	-1448
O=O	+497		
	+1212		-1448

ΔH = +1212 + (-1448) kJ mol^{-1} = -236 kJ mol^{-1}.
(b) -394 kJ mol^{-1}
(c) The C=O bond energy in the data book is a mean value and is not correct for CO_2. (The C=O bond energy in CO_2 is about 800 kJ mol^{-1}.)

16. (a) $\Delta H_7 = \Delta H_1 + \frac{1}{2}\Delta H_2 + \frac{1}{2}\Delta H_3 + \Delta H_4 + \Delta H_5 + \Delta H_6$.
(b) Enthalpy of formation; -392 kJ mol^{-1}.
(c)

ΔH	Name of enthalpy change	Value/kJ mol^{-1}
ΔH_1	Sublimation	+90
ΔH_2	Vapourisation	+20
ΔH_3	Dissociation	+194
ΔH_4	Ionisation	+425

(d) $Br(g) + e^- \rightarrow Br^-(g)$.
(e) ΔH_6 = -672 kJ mol^{-1}.

17. Find ΔH for $\frac{1}{2}N_2(g) + 1\frac{1}{2}F_2(g) \rightarrow NF_3(g)$.
Break $\frac{1}{2}$N≡N and $1\frac{1}{2}$F–F. Make 3N–F. ΔH_f = -58 kJ mol^{-1}. NF_3 is exothermic.

18. (a) (i) ΔH = -231.5 kJ mol^{-1}. (ii) Because the product is steam; it must be water.
(b) $\Delta H_f(H_2O)$ =[(-231.5) + (-47)] kJ mol^{-1} = -278.5 kJ mol^{-1}. (The accepted value is -286 kJ mol^{-1}. The difference is due to the use of mean bond energies.)

19. (a) z is the heat of formation, ΔH_f.
(b) w is +2145 kJ mol^{-1}; x is +1308 kJ mol^{-1}; y is -3428 kJ mol^{-1} (formation of six C–H, one C–C and one C=C).
$z = w + x + y$ = [(+2145) + (+1308) + (-3428)] kJ mol^{-1} = +25 kJ mol^{-1}.

20. (a) y = +232.5 kJ mol^{-1}.
(b) -1111 kJ mol^{-1} = $x + y$ + (3 x -645) kJ mol^{-1}. Hence x = +591.5 kJ mol^{-1}.
(c) z = 591.5 + 232.5 = 824. (N.B. z is a number only.)

21.

Bond breaking	ΔH/kJ mol^{-1}	Bond making	ΔH/kJ mol^{-1}
Be(s→g)	+321	2 x Be–Cl	-778
Cl–Cl	+243	$BeCl_2$(g→s)	$-x$
	+564		$-778 - x$

ΔH = (+564) + (-778 − x) = -512 kJ mol^{-1}.
Hence x = +298. The heat of sublimation of $BeCl_2$(s) is therefore +298 kJ mol^{-1}.

22. (a) The reaction with HBr is addition.
(b) (i) 60°. (ii) About 109°. (iii) Bond strain.
(c) (i) Smaller. (ii) The ring opens easily on addition so the bonds are weaker.
(d) 337 kJ mol^{-1}.
(e) Break one C–C bond to open the ring (= x kJ mol^{-1}) and an H–Br. Make one C–H and one C–Br. Add numbers: x + 366 − 414 − 280 = -3 from which x = 325. Hence the C–C bond energy in cyclopropane is +325 kJ mol^{-1} (smaller than the mean).

23. (a) $\Delta H_1 = \Delta H_2 + \Delta H_3$. (b) $\Delta T_1 = \Delta T_2 + \Delta T_3$.
    (c) ΔT_1 = 22.5 °C, ΔT_2 = 10.1 °C, ΔT_3 = 12.4 °C.
    $\Delta T_2 + \Delta T_3$ = (10.1 + 12.4) °C = 22.5 °C = ΔT_1, which confirms Hess's Law.
24. x = −10 kJ mol^{-1}; y = −270 kJ mol^{-1}; z = +170 kJ mol^{-1}.
25. (a) Combustion.
    (b) It is impossible to burn methane in limited oxygen so that carbon monoxide only is obtained. Some carbon dioxide and free carbon will be obtained.
    (c) (i) $\Delta H_1 = \Delta H_2 + \Delta H_3$. (ii) Hess's law. The enthalpy change for a reaction is the same whatever the route. (iii) ΔH_2 = −599 kJ mol^{-1}.
26. ΔH = −36 kJ mol^{-1}.
27. (a) 1: sublimation; 2: ionisation; 3: hydration. (b) ΔH = +205 kJ mol^{-1}.
28. (a) Elements are assumed to have zero enthalpies (i.e., zero enthalpies of formation).
    (b) ΔH_1: dissociation, +949 kJ mol^{-1}; ΔH_2: dissociation, +436 kJ mol^{-1}; ΔH_3: formation, −46 kJ mol^{-1}.
    (c) (i) $2\Delta H_3 = \Delta H_1 + 3\Delta H_2 + 6x$. (ii) ΔH(formation of N–H bond) = −391.5 kJ mol^{-1}.
29. (a)

Enthalpy /kJ mol^{-1}
638 — 2H(g) + Se(g)
ΔH_2
$2x$
Not to scale.
436 — 2H(g) + Se(s)
86 — $H_2Se(g)$
ΔH_1
ΔH_f
0 — $H_2(g)$ + Se(s)

(b) Let x = H–Se bond energy.
$\Delta H_1 + \Delta H_2 + 2x = \Delta H_f$
x = −276 kJ mol^{-1}.

30. ΔH_f = −251 kJ mol^{-1}.
31. ΔH_c = −727 kJ mol^{-1}.
32. 464 kJ mol^{-1}.
33. (a) 431 kJ mol^{-1}.
    (b) ΔH_f = −91.5 kJ mol^{-1}.
34. (a) −203 kJ mol^{-1}. (b) (i) $\frac{1}{2}I_2(s) + \frac{1}{2}Cl_2(g) \rightarrow ICl(g)$. (ii) ΔH_f = +17.5 kJ mol^{-1}.
35. ΔH (electron affinity) = −381.5 kJ mol^{-1}.
36. (a) Exo. (b) Endo. (c) Either. (d) Endo. (e) Either. (f) Endo. (g) Endo. (h) Either. (i) Endo. (j) Exo. (k) Exo. (l) Endo. (m) Endo.

Unit 7

1. (a) (i) 1 mol; (ii) 3/4 mol; (iii) 3/4 mol.
   (b) (i) Time 0; (ii) Time 0; (iii) Time x and thereafter.
   (c) (i) $(1/2)/(1\frac{1}{2}) = 1/3$. (ii) The equilibrium mixture. (iii) By adding a catalyst.
2. (a) $CH_3COOH + C_2H_5OH \rightarrow CH_3COOC_2H_5 + H_2O$.
   (b) Reversible reaction.
   (c) The composition does not change but forward and reverse reactions continue at equal rate.
   (d) The ester is immiscible with water while the acid and alcohol are miscible.
   (e) (i) Water would shift the equilibrium to the left, decreasing the amount of ester.
   (ii) Because the action of water (hydrolysis) is very slow.
3. Zn(s) → $Zn^{2+}(aq)$
   Zn(s) ← $Zn^{2+}(aq)$
   Radioactive zinc atoms go into solution as zinc ions at the same rate as zinc ions revert to zinc atoms.
4. (a) Methane and steam. (b) The forward rate. (c) They are equal.

(d) (i) The backward rate. (ii) Exothermic. (iii) Since r_b is increased more than r_f, equilibrium shifts to the left. The reverse reaction therefore absorbs heat and so the forward reaction is exothermic. (iv) The equilibrium at z contains more reactant and less product than at y.

5. An equilibrium exists between undissolved salt and the dissolved salt in the saturated solution: $NaCl(s) \rightleftharpoons NaCl(aq)$. Given enough time the radioactive species becomes uniformly distributed between solid and solution.
6. An equilibrium exists between the crystal and the saturated solution: the crystal dissolves at the same rate as solute crystallises. However, the solute that crystallises does not deposit where the crystal dissolves. So the crystal changes its shape.
7. (a) The iodine spreads itself between the two solvents.
   (b) (i) The distribution (partition) of iodine is the same in each tube.
   (ii) The tubes have the same appearance.
8. (a) (i) Equilibrium % decreases; (ii) Equilibrium % increases.
   (b) 24%.
   (c) Ammonia is removed before equilibrium is attained because this is faster and therefore more economical.
   (d) Advantage: the reaction is faster. Disadvantage: the % conversion is smaller.
   (e) (i) To withstand the very high pressures used in the process. (ii) High cost.
9. (a) The decomposition lacks activation energy.
   (b) (i) $N_2(g) + O_2(g) \rightleftharpoons 2NO(g)$; (ii) Endothermic; (iii) Shifts the equilibrium to the right; (iv) No effect.
   (c) The nitrogen monoxide produced by lightning becomes nitrogen dioxide and then nitric acid. This reacts with minerals in the soil to give nitrates, which plants use as nutrients.
10. B.
11. The mixture is not at equilibrium. The gases should react but lack of activation energy at room temperature prevents this. The insertion of a catalyst (e.g., platinum) allows the reaction to occur (which it usually does explosively) to the equilibrium state (which is water with little residual hydrogen and oxygen).
12. (a) False. (b) False. (c) True. (d) False (increases forward and reverse rates equally).
    (e) False. (f) True. (g) True.
13. (a) (i) Curve 2; (ii) A catalyst lowers the height of the curve, which is the activation energy.
    (b) It increases forward and reverse rates equally but does not affect the composition.
    (c) (i) It would shift to the left. (ii) ΔH (from the diagram) is negative: the forward reaction is exothermic. Hence raising the temperature shifts the equilibrium in the endothermic direction towards reactants.
14. Hydrogen chloride supplies chloride ion, which disturbs the equilibrium

$$K^+Cl^-(s) \rightleftharpoons K^+(aq) + Cl^-(aq)$$

by shifting it to the left. Solid potassium chloride is therefore precipitated from the saturated solution, hence the cloudiness.

15. The equilibrium is: $2H_2(g) + O_2(g) \rightleftharpoons 2H_2O(l)$. The forward reaction is exothermic. Raising the temperature shifts the equilibrium to the left and increases the number. Increasing the pressure shifts the equilibrium to the right and decreases the number.

16. (a) (i) ΔH_1 is positive; (ii) ΔH_2 is negative.
    (b) Increased pressure shifts the equilibrium to the left, increasing the proportion of N_2O_4. Of itself, this would make the colour paler. Increased pressure without a shift would make the colour darker because the same gases are confined to a smaller volume. The opposing effects result in the new equilibrium being slightly darker!
17. (a) (i) Yellow-brown; (ii) Yellow-brown.
    (b) Less brown, more yellow.
    (c) To the right.
18. (a) It has two steps instead of one. It has a valueless product that would be difficult to dispose of.
    (b) A mixture of ethene and oxygen (or air) is explosive. (This reaction has to be very carefully controlled.)
19. Raw material-oriented = Rmo; product-oriented = Po. (i) Rmo (usually); (ii) Rmo; (iii) Po; (iv) Rmo; (v) Po; (vi) Rmo; (vii) Rmo; (viii) Po; (ix) Rmo; (x) Rmo; (xi) Rmo; (xii) Rmo.
20. (a) The equation requires 1 volume N_2 to 3 volumes H_2, i.e., 25% N_2 and 75% H_2, which is in approximate agreement with the values in the table.
    (b) (i) Methane, naphtha and steam; (ii) Air; (iii) Air.
    (c) (i) It removes carbon monoxide. (ii) Carbon monoxide would poison the catalyst.
    (d) (i) Iron. (ii) To increase surface area.
    (e) The equilibrium %, though higher, would take longer to form. Greater economy is achieved by taking less sooner and recycling the unreacted gases.
    (f)

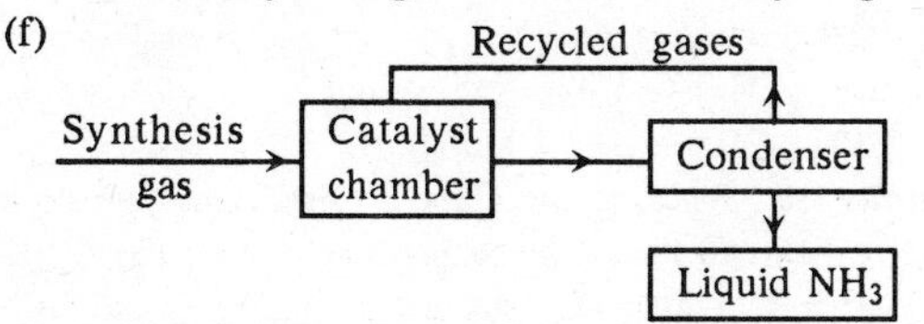

    (g) (i) Because it is unreactive.
    (ii) Shifts the equilibrium to the left.
    (iii) Dilution with argon has the same effect as reducing the pressure of the reacting gases.
21. (a) Raw materials are materials from which products are manufactured. (Primary raw materials are coal, oil, gas, ore, minerals in general, grain, water, air.) The product of one industry may be the 'raw material' of another (e.g., crude oil gives ethene, ethene gives polythene, polythene gives kitchen ware).
    (b) Industrial plant is subject to wear and tear and has to be repaired or replaced. Or it may become obsolete and have to be scrapped altogether. (There are two main reasons for obsolescence: a better way of producing the same product is discovered; the product is superseded by a better product.)
    (c) Skilled and unskilled labour is required. Costs vary depending on manning levels and the nature of the work. Research and development can increase costs considerably.
    (d) Utilities depend on the nature of the process and include electricity, gas, oil, water and steam. The use of waste heat exchangers and recycling energy within the plant can keep these costs at a minimum.
    (e) Overheads are indirect costs incurred by the plant. Examples: accident and health care, fire services, works security, canteen facilities.
22. (a) Because its removal shifts the equilibrium to the left and leaves OH^- in excess.

(b) It turns red in alkaline solution.

(c) (i) Hydrogen; (ii) Apply a burning taper — the gas pops when mixed with air.

(d) (i) $Ca(s) + 2H^+(aq) \rightarrow Ca^{2+}(aq) + H_2(g)$; (ii) Displacement, redox.

(e) Hydrogen ion is rapidly replaced by ionisation of water molecules.

(f) (i) Hydrogen ion; (ii) $2H^+(aq) + 2e^- \rightarrow H_2(g)$.

(g) (i) The equilibrium shifts to the left.

(ii) The phenolphthalein turns red proving that excess OH^- ions are produced; these ions move to the positive electrode, which is what we expect of negative ions.

23. (a) 1. (b) 4. (c) 2. (d) -1.
24. (i) 2; (ii) 7; (iii) 13; (iv) 0; (v) -1.
25. (a) 8. (b) 7.
26. 7.
27. 7. The H^+ produced by ethanol is only $^1/_{100}$th that already present in the water and so it can be ignored. (The answer is not 9!)
28. (a) These data are subject to experimental error and so a 'best line' is drawn through the points. The best lines here look like straight lines.

(b) (i) Hydrochloric acid: pH = 1; ethanoic acid: pH = 3.5. (ii) Hydrochloric acid is strong and is fully ionised. Ethanoic acid is weak and is only slightly ionised.

(c) 10^{-5} mol L^{-1}.

(d) By the seventh dilution, the $[H^+]$ due to the acid is 10^{-7} mol L^{-1}, the same as that due to water itself. Further dilution therefore cannot decrease $[H^+]$ below this value and so the pH is 7 thereafter.

29. (i) $[OH^-] = 10^{-11}$ mol L^{-1}; (ii) $[OH^-] = 10^{-5}$ mol L^{-1}.
30. (i) $[OH^-] = 3.3 \times 10^{-9}$ mol L^{-1}; (ii) $[OH^-] = 1.4 \times 10^{-7}$ mol L^{-1}.
31. (i) $[H^+] = 1.7 \times 10^{-2}$ mol L^{-1}; $[H^+] = 1.1 \times 10^{-13}$ mol L^{-1}.
32. (a) (i) Shifts to the right. (ii) Endothermic. When temperature is raised an equilibrium shifts in the direction that absorbs heat (to lower the temperature — Le Chatelier).

(b) (i) 7; (ii) 6.5. ($10 \times 10^{-14} = 10^{-13}$. $\sqrt{10^{-13}} = 10^{-6.5}$.)

(c) Because it is the nearest temperature to ambient temperature that gives a whole number for the pH of water.

33. (a) $HCOOH(aq) \rightleftharpoons HCOO^-(aq) + H^+(aq)$.

(b) Strength increases with dilution. (All acids are said to be equally strong at infinite dilution.)

(c) pH lies between 3 and 4. If it were fully ionised its pH would be 3. But it is only 45% ionised. So its pH is higher, but not as high as 4. 3.5 is a reasonable estimate. (The calculated value is 3.35.)

34. (a)

$[H^+]$/mol L^{-1}	y	pH
1×10^{-8}	0	8
2×10^{-8}	0.35	7.65
5×10^{-8}	0.72	7.28
6×10^{-8}	0.80	7.20
9×10^{-8}	0.95	7.05
10×10^{-8}	1	7

(b) pH = 6.38. ($0.4 \times 10^{-6} = 4 \times 10^{-7}$, from which pH = 7 - 0.62 = 6.38.)

35. (i) 1 mol L^{-1}; (ii) 10^{-2} mol L^{-1}; (iii) 10^{-9} mol L^{-1}; (iv) 10^{-14} mol L^{-1}; (v) $10^{-11.1}$ mol L^{-1} = 7.9 x 10^{-12} mol L^{-1}; (vi) $10^{-5.6}$ mol L^{-1} = 2.5 x 10^{-6} mol L^{-1}; (vii) $10^{-2.9}$ mol L^{-1} = 1.3 x 10^{-3} mol L^{-1}; (viii) $10^{-0.5}$ mol L^{-1} = 3.2 x 10^{-1} mol L^{-1}; (ix) 10 mol L^{-1}; (x) 10^{-15} mol L^{-1}.

36. The hydroxylic hydrogen is δ+ because it is attached to an electronegative oxygen atom, which is δ–. This causes a small proportion of these hydrogen atoms to ionise as positive hydrogen ions. The methyl hydrogens are not charged in this way and remain bonded to the methyl carbon.

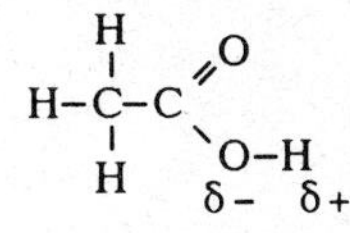

37. (a) 4. (b) 3. (c) 6.

38.

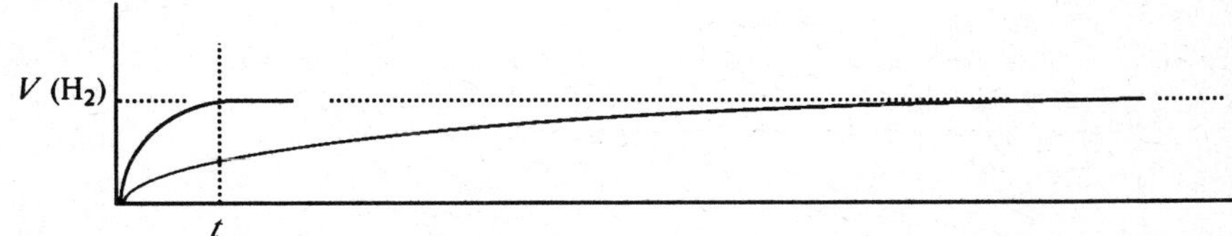

39. (a) (i) Four colours. (ii) Red (0–5), orange (5–6), yellow (6–10), green (10–14).
(b) Because it is also yellow between $6^1/_2$ and 7, which is on the acid side of 7.
(c) (i) Thymolphthalein. (ii) The salt produced in this titration is sodium ethanoate, which has a pH of about 10. Thymolphthalein changes colour at this pH and would therefore indicate when the correct volume of alkali had been added to the acid.
(d) If the solution is red, its pH must be 4 or less. If 4, the original acid must have been pH = 0 (1 mol L^{-1}) in order to account for the 10 000-fold dilution.

40. (a) Same reaction in each cylinder: $Mg(s) + 2H^+(aq) \rightarrow Mg^{2+}(aq) + H_2(g)$.
(b) (i) The reaction with hydrochloric acid is faster.
(ii) Hydrochloric acid is fully ionised (strong); ethanoic acid is slightly ionised (weak). The greater concentration of H^+ in the strong acid allows a faster reaction.
(iii) For a given concentration, the strong acid has a lower pH and a greater conductivity.
(c) (i) The final volumes are the same.
(ii) Each acid supplies the same amount of hydrogen ion for the reaction but the weak acid takes longer to do so.

41. (a) (Mono)chloroethanoic acid.
(b) (i) $CH_2BrCOOH(aq) \rightarrow CH_2BrCOO^-(aq) + H^+(aq)$.
(ii) See answer to Question 36.
(c) Acid 3 (chloroethanoic acid).
(d) (i) Chlorine, bromine and iodine are more electronegative than the hydrogen they replace. There is, as a consequence, a 'knock-on effect' that makes the hydroxyl group more polar and so the halogenated acids are stronger. Electronegativities increase from iodine to chlorine and so acid 1 is weakest and acid 3 strongest.
(ii) Above acid 1; ethanoic acid is weaker than CH_2ICOOH.
(e) $[H^+]$ = 0.026 mol L^{-1} = 2.6 x 10^{-2} mol L^{-1}. pH lies between 1 and 2, nearer 2 than 1. A reasonable estimate is 1.7. (Calculated value is 1.59.)

42. (i) Shifts to the left; (ii) Shifts to the right; (iii) Shifts to the left.

43. (a) A base that is only slightly ionised in solution (less than about 5%).
    (b) $NH_3(aq) \equiv NH_4OH(aq) \rightleftharpoons NH_4^+(aq) + OH^-(aq)$.
    (c) (i) $[OH^-] = 0.01 \times 0.001$ mol $L^{-1} = 10^{-5}$ mol L^{-1}.
    (ii) pOH = 5, therefore pH = 9. (Note that pH + pOH = 14.)
44. (a) (i) Shifts to the left; (ii) Shifts to the right; (iii) Shifts to the right.
    (b) $CH_3NH_2(g) + H_2O(l) \rightleftharpoons CH_3NH_3^+(aq) + OH^-(aq)$.
    (c) The ionisation is greater.
45. (a) $CH_3NH_3^+(aq) + OH^-(aq)$ or CH_3NH_3OH.
    (b) (i) Equimolar means that each solution contains the same amount (in mol) per litre.
    (ii) So that differences in $[OH^-]$ are due only to differences in ionisation and not to differences in concentration.
    (c) (i) Ethylammonium hydroxide is the strongest of the three bases.
    (ii) It has the highest pH and is therefore the most ionised.
46.

Acids		Bases	
Strong	Weak	Strong	Weak
HNO_3	CH_3COOH	KOH	$C_2H_5NH_2$
HI	H_2CO_3	$Ca(OH)_2$	CH_3NH_2
HBr	H_2SO_3	NaOH	NH_3
H_2SO_4	HCN	RbOH	
$HClO_4$	HF		
HCl	HCOOH		
	C_2H_5COOH		
	C_6H_5OH		
	C_2H_5OH		

47. (a) 13. (b) 11. (c) 10.
48. (a) Hydrogen sulphide, H_2S.
    (b) (i) Weak.
    (ii) Hydrolysis gives excess $OH^-(aq)$:
    $$Na_2S(s) + 2H_2O(l) \rightarrow 2Na^+(aq) + 2OH^-(aq) + H_2S(aq).$$
    This reaction occurs because H_2S is weak and its ions join together.
    (c) (i) NaHS. (ii) The acid salt.
    (iii) Hydrolysis of the acid salt does not give as much $OH^-(aq)$:
    $$NaHS(s) + H_2O(l) \rightarrow Na^+(aq) + OH^-(aq) + H_2S(aq).$$
49. (a) Ammonium sulphate is the salt of a strong acid and weak base and is hydrolysed to give excess hydrogen ion, hence the acidity.
    (b) Lime (calcium oxide/hydroxide) or chalk (calcium carbonate) is added.
50. (a) Hydrolysis of ammonium chloride creates excess hydrogen ion:
    $$NH_4Cl(s) + H_2O(l) \rightarrow NH_4OH(aq) + H^+(aq) + Cl^-(aq).$$
    This reaction occurs because ammonium hydroxide is a weak base and so its ions join together.
    (b) Hydrolysis of sodium carbonate creates excess hydroxide ion:
    $$Na_2CO_3(s) + 2H_2O(l) \rightarrow 2Na^+(aq) + 2OH^-(aq) + H_2CO_3(aq).$$
    This reaction occurs because carbonic acid is a weak acid and so its ions join together.

(c) (i) 7.

(ii) Potassium nitrate is dissociated by water:

$$KNO_3(s) + \text{water} \rightarrow K^+(aq) + NO_3^-(aq).$$

Since KOH and HNO_3 are both strong, neither OH^- nor H^+ is removed from the water. The solution therefore remains neutral.

(d)

Type of salt			
Strong acid /strong base	Strong acid /weak base	Weak acid /strong base.	
KCl	NH_4I	$NaHCO_3$	NH_2COONa
NaBr	CH_3NH_3Cl	KF	KCN
Rb_2SO_4	$(NH_3CH_2COOH)Cl$	CH_3COONa	K_2SO_3
$CsNO_3$		Na_2CO_3	HCOONa
$CaSO_4$		C_2H_5ONa	BaF_2
KBr		C_6H_5OK	

51. Ammonium chloride is the salt of a strong acid and a weak base. It is hydrolysed to give excess hydrogen ion. This hydrogen ion neutralises some of the hydroxide ion produced by the ionisation of ammonia. This lowers the pH.
52. Sodium ethanoate is the salt of a weak acid and strong base. It is hydrolysed to give excess hydroxide ion. This hydroxide ion neutralises some of the hydrogen ion produced by the ionisation of ethanoic acid. This raises the pH.
53. (a) Orange. (b) It shifts the equilibrium to the right. (c) Red.

Unit 8

1. $A_r(Zn) = 65.4$.
2. (a) (i) 3. (ii) Mg-24. (b) $A_r(Mg) = 24.3$.
3. B-10: 20%; B-11: 80%.
4. (a) Greater. (b) 62.9 is less than the 63.5 average. The r.a.m. of the other isotope must therefore be greater than 63.5. (c) A_r(Cu-65) = 64.8 (from these data).
5. (i) $^{14}CO_2$, 46 u; (ii) ^3HOH or HO^3H, 20 u; (iii) $CH_3{}^{17}OH$, 33 u.
6. (a) (i) $C_2H_5^+$ and OH^+. (ii) Structure 2. (iii) If structure 2 were correct, these ions would be formed by breakage of the C–O bond. (b) This could be due to either CH_3O^+ from structure 1 or to CH_2OH^+ from structure 2.
7. (a) 2: $^{81}Br^+$; 3: $^{79}Br_2^+$; 4: $^{79}Br^{81}Br^+$; 5: $^{81}Br_2^+$.
   (b) (i) 160 u. (ii) Percentages at 158 u and 162 u are the same and so the average is 160 u.
8. (a) 1: $^{35}Cl^+$; 2: $^{37}Cl^+$; 4: $^{35}Cl^{37}Cl^+$; 5: $^{37}Cl_2^+$.
   (b) Cl-35: 30 atoms at 35 u → 1050 u. Cl-37: 10 atoms at 37 u → 370 u. Mass of 40 atoms = (1050 + 370) u = 1420 u. A_r = 1420/40 = 35.5.
   (c) (i) Isotopic variation produces peaks 1 and 2 for the Cl atom and peaks 3, 4 and 5 for the Cl_2 molecule. (ii) Fragmentation of the Cl_2 molecule produces peaks 1 and 2.
   (d) 17.5. (This ion is deflected as an ion of relative mass 17.5 carrying a single charge.)
9. (a) (i) 6:6 and 7:6 → 6.5:6 = 1.08:1. (ii) 126:83 = 1.52:1. (b) The ratio increases.
10. (a) (i) $^{232}_{90}Th \rightarrow {}^{228}_{88}Ra + {}^{4}_{2}He$; (ii) $^{238}_{92}U \rightarrow {}^{234}_{90}Th + {}^{4}_{2}He$.
    (b) (i) $^{14}_{6}C \rightarrow {}^{14}_{7}N + {}^{0}_{-1}e$; (ii) $^{3}_{1}H \rightarrow {}^{3}_{2}He + {}^{0}_{-1}e$.

(c) (i) $^{40}_{19}K + ^{0}_{-1}e \rightarrow ^{40}_{18}Ar$; (ii) $^{54}_{25}Mn + ^{0}_{-1}e \rightarrow ^{54}_{24}Cr$.

(d) (i) $^{238}_{92}U + ^{1}_{0}n \rightarrow ^{239}_{92}U$; (ii) $^{40}_{18}Ar + ^{1}_{0}n \rightarrow ^{41}_{18}Ar$.

11. (a) Cosmic radiation.
(b) (i) $^{14}_{7}N + ^{1}_{0}n \rightarrow ^{14}_{6}C + ^{1}_{1}H$; (ii) $^{14}_{7}N + ^{1}_{0}n \rightarrow ^{11}_{5}B + ^{4}_{2}He$; (iii) $^{14}_{7}N + ^{1}_{0}n \rightarrow ^{12}_{6}C + ^{3}_{1}H$.

12. (a) $^{24}_{11}Na$. (b) $^{1}_{0}n$. (c) $^{1}_{0}n$. (d) $^{148}_{58}Ce$. (e) $^{2}_{1}H$. (f) $^{17}_{8}O$. (g) $^{1}_{1}H$. (h) $^{0}_{-1}e$.

13. (a) (i) P: protoactinium; Q: uranium; R: thorium. (ii) Uranium. (b) x: β; z: α.

14. (a) About 8 days. (Draw a graph or use 'fraction left = $(^1/_2)^n$'.)
(b) (i) About 22. (ii) Different pattern. (iii) Individual disintegrations occur at random.

15. (i) $^1/_2$; (ii) $^1/_4$; (iii) $^1/_8$; (iv) $^1/_{16}$.

16. (i) 0.71; (ii) 8.9×10^{-16}.

17. (a) True. (b) False. (c) False. (d) True. (e) True.

18. (a) $^{14}CO_2(air) \rightleftharpoons ^{14}CO_2(water) \rightleftharpoons ^{14}CO_2(CaCO_3)$, i.e., exchange of atmospheric $^{14}CO_2$ through ions in solution with the CO_3^{2-} ion in the limestone.
(b) The bonding in these materials is covalent and so carbon exchange is not possible.

19. (a) 5 days. (b) Bi-210.
(c) $^{210}_{83}Bi \rightarrow ^{210}_{84}Po + ^{0}_{-1}e$, $^{210}_{84}Po \rightarrow ^{206}_{82}Pb + ^{4}_{2}He$. (d)

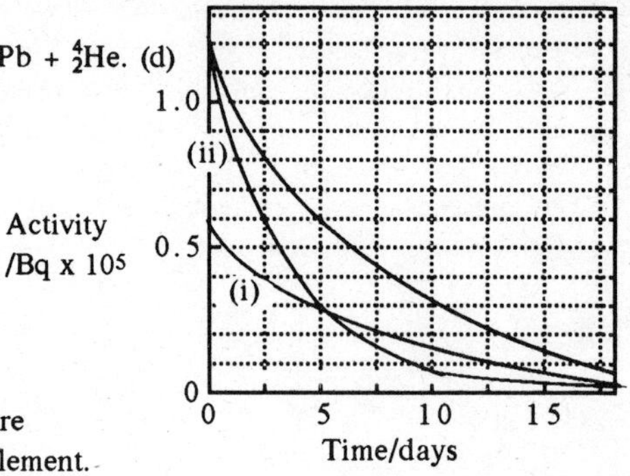

20. (i) 2 mol; (ii) 1.5 g; (iii) 0.625 L;
(iv) A quantity: 7.5×10^7 Bq.

21. (i) 1 mol; (ii) 20 g; (iii) 4000 mL;
(iv) A quantity: 4×10^{11} Bq.

22. (i) 2 $t_{1/2}$; (ii) 3 $t_{1/2}$; (iii) 4 $t_{1/2}$;
(iv) 1 $t_{1/2}$.

23. (a) (i) 4; (ii) Negative; (iii) 0; (iv) $^{4}_{2}He$.
(b) (i) α; (ii) β; (iii) γ.

24. (a) Not repelled by nucleus because they are neutral; give a radioisotope of same element.
(b) (i) Tracer technique. (ii) The radioactive salt acts as a 'marker' for the normal salt.
(c) Electromagnetic radiation of very short wavelength.
(d) There is no blood supply to the graft, which has therefore been rejected.

25. (a) (i) B; (ii) C; (iii) A. (b) (i) C; (ii) A.

26. Two half-lives: 11 400 years old.

27. About 5200 years. (Draw a graph or use 'fraction left = $(^1/_2)^n$'.)

28. (a) The more stable a nucleus, the less energy it contains per nucleon.
(b) The conversion of one element into another.
(c) A: fusion; B: fission.
(d) The problem of maintaining and containing the extremely high temperatures needed.
(e) (i) U-235 or Pu-239. (ii) Each of these nuclei gains a neutron and splits into two smaller nuclei, releasing energy and fresh neutrons. These neutrons repeat the process to give a steady supply of energy. (iii) Many. See, e.g., question 12(d).

29. (a) (i) $\Delta m = -0.0256 \times 10^{-3}$ kg. (ii) $\Delta E = -2.3 \times 10^{12}$ J. (iii) Finite fuels like coal will be used up; there is enough deuterium in water to last a very long time.
(iv) $^{2}_{1}H + ^{3}_{1}H \rightarrow ^{4}_{2}He + ^{1}_{0}n$. (b) $\Delta H = -572000$ J mol^{-1}.
(c) (i) Deuterium. (ii) 4×10^6. (iii) Each uses the same mass of fuel.

30. (a) 560 years. (b) 9.5×10^{-7}.

Printed by Martin's The Printers Ltd., Berwick upon Tweed